TWO MEN WENT TO MOW

CLIVE GRAVETT

First published by Unicorn
an imprint of Unicorn Publishing Group LLP, 2018
5 Newburgh Street
London W1F 7RG
www.unicornpublishing.org

Every effort has been made to trace copyright holders and to
obtain their permission for the use of copyright material. The
publisher apologises for any errors or omissions in the above list
and would be grateful if notified of any corrections that should
be incorporated in future reprints or editions of this book.

10 9 8 7 6 5 4 3 2 1

ISBN 978-1-911604-47-1

Design by Susana Cardona

Printed and bound by Fine Tone Ltd

CONTENTS

ACKNOWLEDGEMENTS

Museum of English Rural Life, Reading, Berks
For use of their archives and support from their staff both
in person and by email. Also for images pages 55, 70 & 76.

Museum in the Park, Stroud, Glos.

Ron Geesin, author of *The Adjustable Spanner, History,
origins and development to 1970.*

Matthew Schneiderman, USA, author of *Edwin Budding
and His Pepperbox: A 21st Century Update.*

John Pease, author of *The History of Thomas Green & Son Ltd.*

James B. Ricci, USA, author of *Hand, Horse, & Motor, The
development of the lawn mower industry in the United States.*

Royal Horticultural Society, Lindley library.

Joseph Paxton, Charles Wentworth Dilke & John Lindley
for founding the *Gardeners Chronicle* in 1841.

Sheffield Postcard Co.

Repton School, Derbyshire.

Brian Bell, author of *Ransomes, Sims & Jefferies,
A history of their products.*

Jack Czislowski, Tasmania, Australia.

Evie Mead for being so photogenic and appearing in nearly
as many images as the author!

Colin Stone & Celia Boyle for their Budding photography.

Alex & John Budding for sharing their family with me.

MISS MARIE STUDHOLME.

FOREWORD

I first met the author two years ago, when I was asked to open the South Downs Heritage Centre and Museum of Gardening, I was completely overwhelmed by his enthusiasm and energy.

To most people a lawn mower means work, but to him it is a passion which he is happy to share.

I was fascinated to read his introduction to the land and lawn mower, because I had a very similar start in that my father and I worked together on the allotment. From an early age I had a gardening round, which involved among other duties, cutting grass using the same machines as the author.

At school I spent more time in the corridor than in the classroom. I was sent to see the headmaster who asked me if I had any interests – rather than just being the school jester. I told him that my great love was gardening, to which he told me to report to the caretaker, who amongst his many duties was maintaining the sports field, which included two football pitches and a cricket table.

It is amazing to see how Clive and I have taken very similar paths in our lives.

This book contains many interesting facts which clearly illustrate the author's passion and dedication to the subject, together with a unique collection of period images. Most people regard cutting the lawn as being like housework, a chore rather than a pleasure.

I believe that reading this book will change a lot of peoples' attitudes, particularly the unbelievable impact that grass and the mower have on sport.

Professionals and keen amateurs will always blame the surface they play on rather than their own inability to adapt to the conditions of the prepared surface.

I was impressed with the detailed research that the author undertook on every aspect of the book and his hero Edwin Beard Budding; much of the material being published for the first time.

Once again our paths crossed because I am very involved with a gardening charity called 'Perennial', which was formed 175 years ago by Charles Dickens to help professional gardeners in need. Today the charity caters for all ages involved with horticulture.

The author has dedicated his incredible enthusiasm to his charity, The Budding Foundation, raising funds to support young people in need.

I strongly recommend this book on a subject we all take for granted.

The only thing I did not discover: does the author have time to cut his own lawn!!

Jim Buttress
RHS Show Judge & holder of the RHS Victoria Medal of Honour

August 2018

INTRODUCTION

In this book I explore the passion of two men, myself and Edwin Beard Budding (1796-1846), both born to a farming family and no doubt sharing a similar relationship with the land during childhood. Our common denominator being the lawn mower!

My interest in all things mechanical came at an early age, my father Denis on a farm worker's wage, was always making and repairing various items to assist our frugal family life. At the age of four I spent many happy hours on his lap helping to drive a grey Fergie tractor and, on reaching seven, I was allowed to drive by myself, but remember having to stand as all my weight was needed to operate the clutch and brakes.

The first encounter with a mower was helping the gamekeeper who lived next door to mow the orchard with an Oxford Allen Scythe, a beast of a machine for a young lad. If you can imagine hanging on the tail of a crocodile whilst it pulled you at speed chomping its way through a reed bed, that's what it felt like!

On reaching ten years of age in 1963 my dad purchased a Suffolk Colt petrol engine mower. I remember many happy years mowing the lawn for him and indeed my imaginary football pitch between the flower beds where I re-enacted the goals scored in

My Father with grandchildren

the 1966 World Cup Final. The goal posts being two rather thorny standard roses that were responsible for puncturing many of my plastic footballs, '*whilst they may have thought it was all over* ', Dinky our corgi dog made good use of the deflated balls spending hours carrying them around the garden!

In my early teens to earn some pocket money I would mow the lawn of my father's boss, John Hunter, with his Suffolk Punch mower, a great treat for me, especially the tea and cake provided by his housekeeper, although I always found his post mow inspection quite daunting.

In my later teenage years, I continued with my interest in mechanics, tinkering/rebuilding my Ariel Leader motorcycle and later Ford Anglias and Escorts.

Whilst I could never see myself doing any other job than farming or horticulture, my mother Edna's wish was for me to get a career and not to end up in a tied property on a minimal wage as they had endured for many years. So at the age of sixteen with a push from mother I entered the world of banking!

Thirty-five years later I was to return to my true vocation, harboured for many years but practised as a serious hobby during my financial sentence; later chapters reveal all.

This book pulls together some interesting facts and insights: where the lawn came from, the mower's creation and its massive impact on sport and social history.

As for my hero Edwin Beard Budding, or should I say the first man who went to mow, he is of course worthy of his own chapter.

Read or should I say 'Mow On'.

BIRTH OF THE
LAWN MOWER

GRASS

Grass has been described as the great motor or engine of the world; with over 11,000 species some only reaching a few millimetres to the mighty bamboo Dendrocalamus Giganteus reaching a massive 42 metres (*equivalent height to a fourteen storey building*)!

Equivalent height to a 14 storey building!

Did you know? Rice, wheat and maize are all grasses and together supply nearly 50 per cent of all the calories consumed by the entire human population.

Thomas Browne, 17th century thinker and polymath, said...

> 'All flesh is grass, is not only metaphorically,
> but literally, true; for all those creatures we behold are but
> the herbs of the field, digested into flesh in them, or more remotely
> carnified in ourselves.'

Remember when next having a barbecue on the lawn... the cow eats grass and makes cheese, the steers and heifers eat grass and become beef, we toast our cheese and barbeque our burgers and wrap them in a bread roll, made of grass!

Grasses evolved around 70 million years ago when forests were the dominant land vegetation; grasses had three weapons to take on the trees.

1) 'Fire' they were well adapted to survive burning as their centre of growth was very low down and could survive as fire passed above.

2) 'Grazing' they could survive being grazed and trampled by animals, whereas young trees could not survive being browsed by them.

3) 'Human beings' grasses enlisted humans, for example, maize cannot survive without human intervention as it is unable to disperse seeds. It is therefore dependant on humans as are many other crop plants, but then we are unable to survive without them!

THE LAWN

Pliny the Younger

Who invented the lawn?

Some credit Pliny the Younger, a lawyer, author and magistrate of Ancient Rome, he apparently describes a lawn around his villa on Lake Como in his many letters written around AD100. However, the translation of his letters is unclear and thought to refer to plants other than grass, possibly moss or another form of vegetation.

We know that the word 'Laune' was first used around 1540 and referred to 'a glade, open space in a forest or between woods, later that century becoming "Lawn"'.

Albertus Magnus

I personally credit Albertus Magnus a 13th-century Swabian nobleman turned Dominican Friar, he produced the first gardening book in 1260 *De Vegetabilis*. An extract from his book states ... '*The ground should be cleared of weeds, flooded with boiling water and laid with turves which should be beaten down with broad mallets & trodden, then the grass may spring forth and closely cover the surface like a green cloth*'.

To me that is the creation of a lawn and very similar to how turf is laid 750 years on.

Due to his contributions to natural philosophy, the plant species *Alberta magna* was named after him (from South Africa better known as The Natal Flame Bush), and also in 1993 the asteroid 20006 was given his name.

There is a story that an American visitor once went to Cambridge, and standing on the edge of the magnificent lawn between Kings College Chapel and the River Cam asked the groundsman how it was made.

'Well Sir, said the man on the mower, you cut it and you roll it, and cut it and roll it, and in a hundred years or so you get it like this.'

So, let's move on to see how the early lawns were tended.

SHEEP AND SCYTHES

Sheep can be credited as being the first mowers; one of the first animals domesticated by humankind in around 10,000BC, they have been used over the centuries to mow grass.

On large country estates the extensive lawns in front of the main house were divided by a 'Ha Ha' which can most easily be described as a ditch or trench, this gave an uninterrupted view and illusion of openness when viewed from the main house. The lawn close to the building was cut by scything with that on the other side of the Ha Ha being mown by sheep.

The first organised game of cricket is said to have taken place in 1550 with the MCC being formed in 1787, perhaps the sport would not have developed as it did without the help of sheep being used to mow the pitch. Robert Grimston,

Cricket game

who remained president of the MCC until his death in 1844, earned a reputation as an extreme reactionary as far as ground maintenance was concerned, preferring sheep for keeping the grass down and resisting the advent of new-fangled machinery even at this relatively late date.

Woodrow-Wilson 28th President of the United States 1913–1921 utilised a flock of forty-eight sheep to mow the White House lawns during World War I saving manpower by cutting the grass and raised $52,823 for the Red Cross through an auction of their wool.

The 'Slips' on a cricket field are so called due to the fielders often slipping on the sheep droppings.

FACT

15th Century man with scythe

The scythe was invented in about 500BC and appeared in Europe during the twelfth and thirteenth centuries. Initially used mostly for mowing grass, it replaced the sickle as the tool for reaping crops by the 16th century, the scythe allowing the reaper to stand rather than stoop.

A scythe consists of a wooden shaft about 170cm (6ft) long called a *snaith* (modern versions are sometimes made from metal or plastic). The snaith may be straight, or with an 'S' curve, but the more sophisticated versions are curved in three dimensions, allowing the mower to stand more upright. The snaith has either one or two short handles at right angles to it – usually one near the upper end and always another roughly in the middle. A long, curved blade about 60–90cm (24–35in) long is mounted at the lower end, perpendicular to the snaith.

Scythes always have the blade projecting from the left side of the snaith when in use, with the edge towards the mower. In principle a left-handed scythe could be made, but it could not be used together with right-handed scythes in a team of mowers, as the left-handed mower would be mowing in the opposite direction.

Peening anvil, cow horn holder & sharpening stone

A scythe blade is prepared by peening the leading edge, this is simply hammering it on a flat surface to draw out the metal making it almost as thin as paper. After peening, the edge is finished and subsequently maintained by very frequent honing with a whetstone, and peened again as necessary to recover the fineness of the edge.

The scythe man would carry a small hammer, together with a peening anvil and a whetstone in a cow horn holder, all clipped to his belt.

The sharpening stone works better when damp and the cow horn holder would therefore be filled with damp moss to keep the stone wet.

Mowing with a scythe was a highly skilled occupation, better done when the grass was wet, so scythe men often worked unsociable hours cutting the grass before dawn or after dusk. Before the introduction of adjustable handles on the

Preparing for Croquet

scythe, extra leather soles were strapped to the bottom of their shoes thereby being able to retain the perfect stance, but achieve a higher cut.

Did you Know ?

Scythes continued to be used into the early 20th century alongside the lawn mower and many enthusiasts worldwide still keep this skill going as a hobby, with many clubs and indeed the Scythe Association of Britain and Ireland.

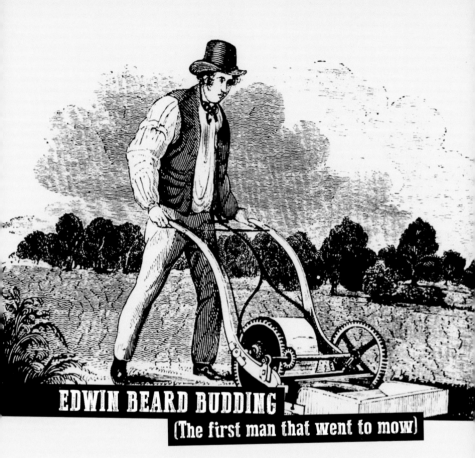

EDWIN BEARD BUDDING
(The first man that went to mow)

Unlike my own childhood there is no definitive record of Edwin Beard Budding. Born in Eastington, Gloucestershire on 25th August 1796, the illegitimate son of Charles Brain Budding a farmer and Mary Beard, hence the Beard-Budding name; siblings elder brother William and younger half-brother Thomas.

To have accomplished what he did in later years, it is hard to believe he probably didn't have any formal education, schools were few and far between with education not being compulsory.

It is possible he attended Sunday school as these gave children basic tuition in reading, writing & arithmetic and it is probable that he worked from an early age.

In 1821 he married Elizabeth Chew; the marriage took place in Hempsted on the outskirts of the city of Gloucester. Three children followed: Frances Anne 1822, Caroline 1824, both baptised in Chalford, and Brice Henry 1830 in Bisley.

Whilst they only moved within their locality we must remember that in the 1820s the only form of transport was horse and cart or walking. Larger towns during this period would have seen the expansion of stage coaches especially following the work of Thomas Telford (1757-1834) who improved road design, drainage and gradients, together with John Macadam (1756-1836) introducing the new road surface 'tarmacadam' or tarmac as it is known today.

In the early 1820s Edwin was described as being a carpenter. It may well be that he had served an apprenticeship in the fairly basic wood working skills associated with the structural elements of building work, but also valuable for constructing and maintaining the mechanisms (such as paddles and wheels) of water-powered mills.

For many centuries this part of Gloucestershire had been a centre for the manufacture of fine quality woollen cloth. The lengths of cloth were traditionally woven in the weavers' homes but were fulled and finished in mills which had been established in the Stroud valleys. The steep valleys provided abundant and efficient water power for the running of the fulling mills. Although the industry was beginning to show signs of decline in the first quarter of the 19th century, it was still the major source of local employment.

Stimulated by the competitiveness of the textile industry, was a burgeoning of technical developments, not only in the ways the wool yarns and the woven cloth were produced, but in the machinery being designed and manufactured locally to facilitate this output. It was in this particular, innovating aspect of the industry, that Edwin Budding became involved. Techniques for improving not only the quality of the cloth, but the speed and ease (and hence lower costs) with which it could be produced, were a constant preoccupation of the mill owners. The technically-minded men they employed were under constant pressure to prototype, develop and refine any new ideas. It was both a highly competitive and therefore also secretive industry.

A major development was the cross-cutting machine for shaving the nap off the shrunk, roughened lengths of 'fulled' woven woollen cloth. This task had previously been carried

Brimscombe Mill

out by skilled men using large scissor-like shears. There were protests that the introduction of such machines would deprive the shear-men of their livelihood, no doubt in a similar vein as the agricultural 'Swing Riots' of the 1830s.

Cross-cutting Machine

In 1815, John Lewis of Brimscombe Mill was the first to put together a number of innovative features in a cast-iron machine for this purpose. The critical element in Lewis's design was the introduction of a set of helical blades attached to a rotating cylinder which passed across the surface of the cloth. Thereafter, many further patented refinements followed. Techniques for achieving a close, smooth pile became increasingly sophisticated. An example of a cross-cutting machine is on display in the Museum in the Park in Stroud, the image seen here clearly shows the helical cutter.

It is understood that Edwin Budding would have worked for/or with John Lewis in the making of these machines, at that time he was described as a 'Machinist' *(A person who machines using hand tools and machine tools to prototype, fabricate or make modifications to a part that is made of metal, plastics, or wood).*

MOWERS DEVELOPMENT

1830 TO 1900

THE LAWN MOWER

It is not clear if (Edwin Beard Budding) was an employee of John Lewis, or contracted to undertake specific tasks, while running his own business. Certainly he must have had access to facilities for working and casting metals. He may have also had workshop premises of his own.

Perhaps on seeing the cross-cutting machine at work, helical cylinder racing down the table shaving the nap, a glance out of the window to see the scythe men cutting the grass around the mill, a moment of inspiration and the two became one and his idea for the lawn mower was born.

Whilst Budding had the genius and skills, he lacked the finance to patent and put the lawn mower into production and he looked to John Ferrabee for help.

Thrupp Mill

When John Ferrabee took on the lease of Thrupp Mill in 1828 he was an iron-founder; he made extensive alterations, which included taking down the dwelling-house and building a foundry. He also removed two of the three waterwheels and their stocks, as steam engines by now provided most of the mill power. The mill under Ferrabee was renamed The Phoenix Iron Works.

Edwin Budding and John Ferrabee signed an agreement relating to the patenting and manufacture of Budding's invention of a new machine for: '*the purpose of cropping or shearing the vegetable surface of Lawns, Grass plats and*

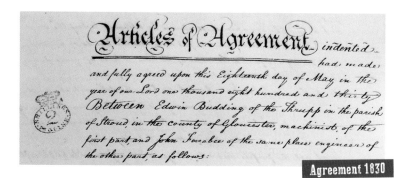

Pleasure Grounds.' Under the agreement, Ferrabee provided the finance for this enterprise and retained rights in the licencing, production and selling of the machine. Budding was enabled to prepare his submission to the Patent Office and was entitled to an equal share in the profits once Ferrabee had recovered his initial outlay. There was a £2,000 penalty should either party default on this agreement. Budding received Letters of Patent for his invention on 31 August 1830.

It is interesting to note that within the patent description Budding states: *'The revolving parts may be made to be driven by endless lines or bands instead of teeth'*. This again shows Budding's genius and foresight as it was not until almost thirty years later in 1859 that Willoughby Green patented the first chain drive mower which revolutionised mower design. (See Thomas Green & Son, Silens Messor page 44)

An accompanying detailed schedule, written in the first person on a large single sheet of parchment, was also prepared. This describes all the carefully lettered components, how they are assembled and how they interact when the machine is operating. Included is an explanation of the gearing system and how the rotating blades cut against a fixed blade.

In describing the attributes of his machine, Budding also made some engaging claims for his invention: 'Grass growing in the shade too weak to stand against a scythe to be cut, may be cut by my machine as closely as required' with the further advantage: 'and the eye will never be offended by those circular scars, inequalities and bare places so commonly made by the best mowers with a scythe and which continue for several days.' Finally, he clearly had notions of the potential market: 'Country Gentlemen may find in using my machine themselves an amusing useful and healthy exercise.'

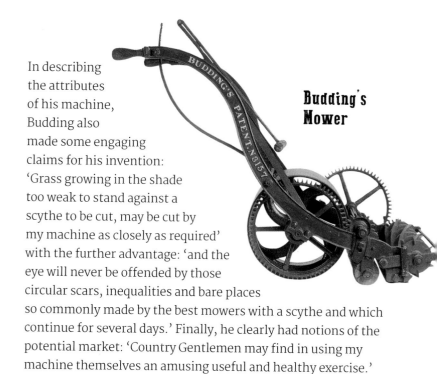

Budding's Mower

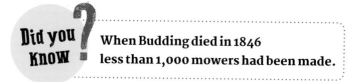

Did you Know?

When Budding died in 1846 less than 1,000 mowers had been made.

There are attractive stories – for which there is, nevertheless, no hard evidence – about the way Budding worked while developing his machine.

It is said that he only tested his prototype at night (presumably to maintain secrecy), but that neighbours in Thrupp complained about the noise it made. This would have been hardly surprising, given its comparatively basic gearing system and the fact that all the interacting moving components parts were made of cast iron. There has been speculation about which areas of grass the prototype was tested on – the lawn of the Ferrabee home, Phoenix House, Thrupp, has been suggested, as have some adjacent meadows.

There must have been not only a working, but also a strong personal bond of trust between Budding and John Ferrabee. In the will the latter signed in 1831, he named Edwin Budding as the sole trustee of his estate (including Phoenix Iron Works), with responsibility for it until Ferrabee's youngest son was twenty-one. But Budding never had to carry out this role as he predeceased John Ferrabee.

Whilst there are no accurate figures for the number of mowers produced at the Phoenix Iron Works, an 1856 Ferrabee advertisement in *The Gardeners Chronicle* states that 5,000 had been sold to date. Edwin Budding died ten years earlier in 1846 so he may have only seen possibly a 1,000 produced.

Budding and Ferrabee did grant a license to J.R. & A. Ransome in 1832, later to become Ransome's, Sims & Jefferies one of the world's major lawn mower manufacturers, they had produced 1,500 Budding pattern lawn mowers by 1852.

The further development of Budding's lawn mower and its impact on the world is continued further on in this book, but let's now take a look at his other inventions…

ADJUSTABLE SPANNER

We are all familiar with the adjustable spanner or wrench, but few are aware that Budding's imaginative brain perceived a way of facilitating frequent adjustments to fit different bolt heads by designing a back rack and worm mechanism or screw adjustment.

Earlier adjustable spanners were of the wedge type and had existed since the 18th century; adjustable spanners were needed as nuts and bolts were cast and not consistent in size.

James Ferrabe Spanne

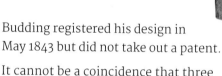

Budding registered his design in May 1843 but did not take out a patent.

It cannot be a coincidence that three months later Richard Clyburn, a Yorkshire engineer who arrived in Dursley in the 1830s, registered a design for an adjustable spanner. Clyburn's was in a slightly different format to Budding's, both forms are still recognisable in those produced today.

Budding Spanner

It is known that Budding and Clyburn would have been working with or in close proximity to each other, could it be they met one evening in the local ale house sharing ideas and decided to register their different designs, perhaps tossing a Victorian penny to decide?

Clyburn Spanne

Wedge Spanner

Budding's spanners were produced by John Ferrabee and later James Ferrabee at the Phoenix Iron Works and whilst the Ferrabees made some improvements after Budding's death, they still promoted 'Ferrabee's Budding Spanner' at the head of their range in their sales catalogues in the 1860s.

JAMES FERRABEE,

FERRABEE'S
PATENT WEDGE SPANNER,
OR, ADJUSTABLE SCREW WRENCH.

The well known and useful Tool called "Budding's Spanner," originally introduced by Mr. Ferrabee, is the most handy form of Screw Wrench for general use; the movable jaw, however, notwithstanding the care taken in the manufacture, is liable to wear, and the spanner in time becomes deranged. To obviate this defect, which is common to all spanners with a sliding jaw, J. F. introduced a wedge between the back of the stem and the movable jaw. On this wedge is fitted the worm which works into the teeth on the stem and moves the jaw up or down as may be desired. The wedge slides with the jaw, and when a strain is applied, the fit of the jaw to the stem and the wedge is so perfect, and the thrust so direct, that the spanner acquires all the strength of a solid one. The stem and upper jaw are made in one piece, and, together with the movable jaw, are of the best forged iron, all accurately fitted by machinery, and case hardened. Thus constructed it is a tool of unrivalled strength, durability, and simplicity, which may be used with confidence under all circumstances and in all situations where an adjusting spanner can be serviceable.

Length	8	10	12	14	15	16	18	20	24	inches.
Range	1	1¼	1½	2	2¼	2½	3	3	3½	,,
Price	6s.	7s. 6d.	9s.	11s.	12s.	13s.	15s.	17s.	20s.	

OTHER LENGTHS MADE TO ORDER.

PORT MILL, BRIMSCOMBE, STROUD, GLOUCESTERSHIRE.

Ferrabee Ad for Budding spanner

Did you Know?

Budding registered his design in 1843 but did not take out a patent.

BUDDING PEPPERBOX PISTOL

As with Budding's other inventions he did not receive recognition for his breakthrough ideas during his lifetime, it took over one hundred years for the 'Gun World' to discover that in 1825 he manufactured what is now considered to be one of the rarest firearms in the world!

In 1962, William Keith Neal (1905–1990) a leading authority on vintage firearms, made the connection with Edwin Beard Budding and the Budding Pepperbox Pistol. Whilst he had been aware of the Budding pistol since 1935, due to no patent having been registered, the inventor remained anonymous.

There were three versions of the pistol all made with five chambers, .30 calibre and all signed 'Budding Maker', it is assumed only around fifty were made.

Pepperbox pistol

Neal stated: *'It is of great interest to note that the nipples set in line with the bore, giving centre fire ignition, was introduced by an Englishman at least six years before Samuel Colts first patent of 1836.'*

Budding Maker Logo

He goes on to say: *'Benjamin Darling, who claimed that he invented the first American revolver, took out his patent in 1836, his is one of the rarest of all American firearms, less than 100 produced. Edwin Budding produced his pepperbox at least six years before Darling. He never patented it and apparently did little to promote it. It is rarer than the Darling and certainly more curious, and deserves a prominent place in firearm history.'*

Edwin Budding produced his pepperbox at least six years before Darling. He never patented it and apparently did little to promote it.

Whilst researching this book I have received a fresh opinion as to why Budding perhaps did not patent his pistol. James Cook, a Birmingham gunmaker, patented a pistol on 30 May 1824 which precisely covers Budding's firearm; there is, however, no historical evidence that tells us whether he knew of Cook's patent or had licence to use it.

One of Cook's specialities was making Gun Walking Canes and I am aware that these types of firearms were also produced by Budding.

Whether or not Budding invented the pistol, he is certainly regarded as an extremely talented engineer of his time.

IMPROVEMENTS IN THE 'CARDING MACHINE'

Wool carding is the cleaning, separating and straightening of the wool fibres being the last stage in the process which prepares a fleece for spinning.

The word 'Carding' is derived from the Latin *carduus* meaning thistle or teasel, as dried vegetable teasels were first used to comb the raw wool.

In 1748, Lewis Paul of Birmingham, England, invented a hand-driven carding machine, the invention was later developed by Richard Arkwright and Samuel Compton.

A young man, George Lister, moved from Yorkshire to Dursley in 1820, and his family in later generations became major producers of heavy machinery and engines for agricultural uses and a wide range of other industries.

George Lister and Edwin Budding worked together in obtaining a patent in 1843 for revolutionary improvements to the carding machine, many of these are still evident in today's machines.

The improvements were extremely complex with the patent document registered on 15 June 1843 running to over 2,200 words.

In the following generations, the Listers became major producers of heavy machinery and engines for agricultural uses and a wide range of other industries, with George Lister's grandson, George, being instrumental in writing letters in the late 1940s telling the story of Budding as he saw it and providing documents to Stroud museum.

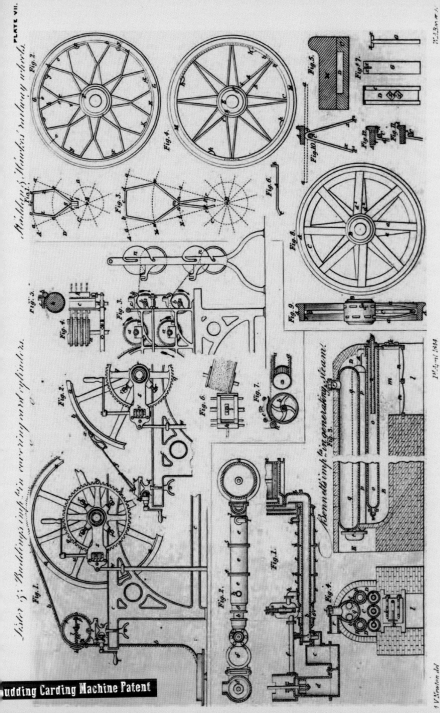

PLATE VII.

Building's, Hunter's railway wheels.

Fig. 2

Fig. 4

Fig. 1

Fig. 3

Fig. 6

Fig. 5

Fig. 7

Fig. 10

Fig. 8

Fig. 9

Leier & Buildings improved carding and spinning cylinder &c.

Fig. 1

Fig. 2

Fig. 3

Fig. 4

Fig. 6

Fig. 7

Kennedy's improved apparatus for generating steam.

Fig. 1

Fig. 2

Fig. 3

Fig. 4

1ˢᵗ April 1844.

J. V. Newton del.

Following Edwin Budding's death in 1846, John Ferrabee continued to produce Budding's patent mowers together with the adjustable spanner until his retirement in 1852. His sons, James and Henry, took over the business but their partnership only lasted three years after which James continued alone.

Patents for the improvements of both the mower and adjustable spanner were applied for in the late 1850s together with improvements to steam engines.

Competition in the production of lawn mowers was increasing with Thomas Green & Son together with Shanks of Arbroath taking a greater share of the market, James ceased production at the Phoenix Iron Works in 1863.

Ferrabee Mowers

HAND MACHINE.—*One Inch Scale.*

HORSE MACHINE.—*Half-Inch Scale.*

Ferrabee Mower. One of the few known to still exist.

SHANKS' PONY

William Fullerton Lindsay Carnegie (1788–1860) of Arbroath Scotland played a prominent part in the development of Scottish industry and public institutions. He was closely involved with the construction of the Arbroath to Forfar railway.

He had experience of Budding's lawn mower but with a lawn of around 2.5 acres he found it inadequate. In discovering Budding's patent had been taken out for England only he employed, in his words 'a very ingenious mechanic in my neighbourhood Mr Shanks of Arbroath to construct a lawn mower for my purpose'.

Alexander Shanks (1801–1845) appears to have shared Edwin Budding's inventiveness, registering his first patent in 1834 for improvements in machinery for preparing and dressing hemp and other fibrous substances.

The first mower made by Shanks for Mr Carnegie had a cutting width of 27in followed by a second machine with a heavier roller and cutting width of 42in. The larger mower pulled by a pony was able to mow and roll the 2.5 acres in under three hours.

Alexander Shanks was granted a patent on 23 July 1842 with his invention being featured in the *Mechanics Magazine* with credit to Carnegie as follows....

Shanks' Grass-Cutting and Rolling Machine
Mechanics Magazine July 23rd 1842

'The proprietors of lawns and pleasure grounds are greatly indebted to Mr Carnegie for bringing under their notice so cheap and efficient a means of keeping them constantly in order, and to the pleasure it must

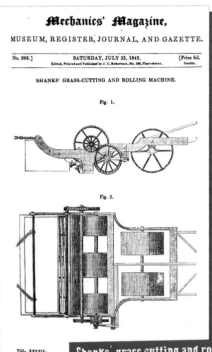

Mechanics' Magazine,

MUSEUM, REGISTER, JOURNAL, AND GAZETTE.

No. 989.]   SATURDAY, JULY 23, 1842.   [Price 6d.
Edited, Printed and Published by J. C. Robertson, No. 166, Fleet-street.   Double.

SHANKS' GRASS-CUTTING AND ROLLING MACHINE.

Fig. 1.

Fig. 2.

VOL. XXXVII.

Shanks' grass cutting and rolling machine - Mechanics' Magazine

give Mr Carnegie to have rendered gentlemen of his class so useful a service, we doubt not, he will soon have to add, the satisfaction of knowing, that his recommendation has been the means of procuring the ingenious inventor, as many orders as he can possibly execute.'

Alexander Shanks & Son progressed to be one of the major lawn mower manufacturers, finally merging this part of their business in 1952 with Charles H. Pugh of Birmingham, makers of Atco lawn mowers. Shanks did continue to manufacture gang mowers until the late 1960s.

Pony and Horse mowers played a vital part in the mowing of large lawns, parks and sports pitches, prior to the invention of the motor mower in the early 1900s.

Alexander Shanks was granted a patent on 23 July 1842 with his invention being featured in the *Mechanics Magazine* with credit to Carnegie.

Shanks pony mower – Balmoral Castle

THESE BOOTS ARE MADE FOR MOWING

With the growth of Horse, Pony and Donkey pulled mowers after their introduction in 1842, there was a need to prevent

Leather boots

injury to the lawn from the animal's hooves/shoes. Advertisements appeared during the 1850s for leather horse boots being sold alongside lawn mowers. In 1859, Thomas Green was advertising sets of four horse boots to accompany their mowers.

The boots were made of leather, being strapped over the animal's hooves with different sizes available for Horse, Pony and Donkey. Some manufacturers offered made-to-measure sizes asking the customer to send a paper pattern of fore and hind hoof to ensure a good fit.

Using a camel to pull a lawn mower was in fact quite logical as they have broad flat feet with leathery pads on the underside of their hooves which spread out when the camel places its feet on the ground thus creating a 'snowshoe effect' and preventing the camel from sinking into the sand.

PATENT "EVERLASTING" SOLID

LAWN MOWING HORSE BOOTS.

These Boots are a first class article, being blocked, sole and upper in one piece, fit better, are more compact and comfortable, and much more durable than any other in use. They retain their shape, and having no sharp edges cannot cut up or mark the lawn and are far superior in every way to the ordinary boot. The Vamp is secured by patent rivets.

When ordering send paper pattern of fore and hind hoof to ensure good fit.

Prices, per Set of Four, for Horse, Cob, or Pony Size, on application.

Pony wearing boots

Camel Mowers - Regent's Park

The accompanying images show camel-pulled lawn mowers in use at Regent's Park, London during the early 1900s.

In 1869, a camel-pulled mower was used to mow the 'Sheep Meadow' a 15-acre lawn in Central Park, New York. The meadow took its name from a flock of Dorset and Southdown sheep that were kept on the meadow from 1864 to 1934.

Camel Mower - Central Park New York

When the sheep failed to do a good job at keeping the grass cut, camel-drawn lawn mowers were used instead.

THE CHAIN EFFECT

If Edwin Beard Budding lived on for a few more years, he would have been the first to have developed the revolutionary chain drive lawn mower. He had stated in his 1830 patent, '*The revolving parts may be made to be driven by endless lines or bands instead of teeth*', however, it was twenty-nine years later in 1859 that Willoughby Green patented the chain drive lawn mower.

The chain drive was a vast improvement with the new mower being called the 'Silens Messor' or Silent Cutter. Advertisements emphasised how much quieter it was when compared with the gear driven machines, often referring to the mower as being noiseless!

Willoughby Green (1837–1870) son of Thomas Green is recorded as working in his father's business as an apprentice at the age of

GREEN'S PATENT SILENS MESSOR.

GREEN'S PATENT SILENS MESSOR, OR NOISELESS LAWN MOWING, COLLECTING, AND ROLLING MACHINES, patent dated December 6th, 1859. THOMAS GREEN in respectfully returning thanks to the Nobility, Gentry, and the Public generally, for the very liberal support he has received for some years past, informs them that (although he has for the last three years, at all the principal AGRICULTURAL AND HORTICULTURAL SHOWS in the kingdom proved the superiority of his Machines over an others, carrying off every prize that has been given, and highly commended by the Judges,) he has taken out this season all entirely NEW PATENT, which he with confidence submits for competition, as it excels all his previous efforts, and overcomes all difficulties.

Greens ad 1859

fourteen. His advancement was rapid as by 1860 at the age of twenty-three he was running the sales operation in London, the main foundry and manufacturing taking place in Leeds. The company name changed in 1863 to Thomas Green & Son, as a consequence of Thomas Green and his son Willoughby entering into a formal partnership. Thomas Green & Son at this time were regarded as the largest lawn mower manufacturers in the United Kingdom.

Willoughby by name and nature was a colourful character; he was a Leeds town councillor and president of Leeds Working Men's Conservative Association, he was also a keen horseman and this sometimes got him into trouble. On one occasion he was fined ten shillings with

Silens Messor

costs for furious driving after a police constable witnessed Willoughby and a friend racing two trotting gigs down the street at an estimated speed of fourteen miles per hour!

Thomas Green remarried in 1869, only four years after the death of his first wife, and it appears that this started a feud between father and son evidenced by newspaper articles describing how a warrant was issued after Willoughby did unlawfully threaten to shoot and murder his father. On the 1 November 1869 the business partnership was ended and on 12 April 1870 Willoughby Green died at the age of thirty-two, cause of death being recorded as 'Disease of the brain'. At the time of his death around 50,000 chain drive mowers had been sold.

Thomas Green & Son's business continued until 1975 when the lawn mower business was acquired by Reekie Engineering of Arbroath having been one of the market leaders in lawn mower sales for almost one hundred years.

FACT

In the 80 years between 1850 & 1930 Thomas Green & Son sold over one million lawn mowers.

Whilst J. R. & A. Ransome were first to manufacture Buddings mower under licence in 1832; their sales did not match Ferrabee in the 1840s or indeed Greens, Samuelson & Shanks during the 1850s, in fact they ceased making mowers for three years from 1858.

Ransomes Automaton Kenilworth-Castle

In 1861, with a name change to Ransome's & Sims, mower production resumed with an improved version of the Budding mower, followed in 1867 by the Automaton which was to dominate Ransome's catalogues for the next thirty years.

Automaton was a curious choice of name, perhaps more appropriate for the robotic mowers used today, in ancient Greek the word meant 'self-moving' or 'self-willed' with today's meaning being along similar lines.

Despite its name, Automaton established the Ransomes a foothold in the lawn mower market with 7,000 being sold in the first six years of production, with minor modifications the Automaton was still available in the early 1930s.

Towards the end of this decade there was a major improvement in mower design whereby the heavy roller which gave the power to drive the cutting cylinder was

dispensed with, being replaced by two much lighter side wheels. Whilst many lawn mower historians credit Follows & Bate of Manchester in inventing the first sidewheel mower, the 'Climax' with their patent dated 31 March 1869, they were in fact beaten by Everett G. Passmore from Philadelphia, USA, who registered a similar patent five weeks prior on 23 February 1869.

Climax – a rare example

Passmore had established a business with John H. Graham & Samuel Emlen in 1858 and according to their 1865 catalogue were wholesale/retail dealers in agricultural and horticultural machinery. His design appears to be the one adopted by most manufacturers worldwide with the simple wooden 'T' handle, whereas Follows & Bate used the traditional double cast handles seen on the earlier roller mowers.

Philadelphia Mower

FOLLOWS & BATE'S
PATENT "CLIMAX" BACK-DELIVERY
LAWN MOWER FOR THE MILLION.
Price 25s.

THE TIMES, *December* 10, 1869.

"FOLLOWS & BATE, of Manchester, bring out the latest little wonder in Lawn Mowers, which is a machine cutting only 6 inches wide, being propelled with surprisingly little force, and costing almost a fractional price as compared with the large machines."

These Machines having no roller in front of the knives, cut LONG or short Grass just as it grows, do not miss the bents, and never choke, however wet the Grass may be.

They are specially applicable for Slopes and Steep Embankments, and are the only Lawn Mowers that can be used effectually with or without the Box.

Between **3000** and **4000** of the "CLIMAX" have been Sold this season. Every Machine is warranted, and a trial allowed.

CATALOGUES, with Testimonials and full particulars of other sizes, on application to any respectable Ironmonger or Dealer in Horticultural Machinery, or sent Post Free from the PATENTEES and MANUFACTURERS,

FOLLOWS AND BATE, MANCHESTER.

MOWER FRENZY IN THE USA

The 1870s saw an explosion of new lawn mower manufacturers in the USA, some following on from the invention of the sidewheel mower; others further developing their traditional roller mowers.

It would appear the first lawn mower manufactured in the USA and imported into England was the Archimedean in 1869, designed by Amariah Hills (1820-1897) and manufactured by Hills Archimedean Lawn Mower Company Inc.

Aaron W.C. Williams travelled to the UK in 1869 to set up Williams & Co, also later in France, Williams exhibited the Archimedean at the International Horticultural Exhibition in Hamburg in September 1869, an event that attracted over 60,000 on one particular day. Whilst some adverting indicated that Williams supplied mowers manufactured in London & Paris, it is thought that they were only manufactured in Hartford, Connecticut and imported for assembly in England.

The company were frequent advertisers in the *Gardeners Chronicle* also as were other American companies presenting colourful artistic trade cards and posters. The adverting certainly seemed to pay dividends with a claim that 16,000 mowers had been sold by 1873.

The name Archimedean referred to the mower's solid cast-iron cutting cylinder, resembling an Archimedean screw.

FACT

USA Trade cards

To compete with the stiff competition from the USA, Follows & Bate introduced the 'Anglo American' mower in 1871; unlike their Climax this was a side-wheel mower made in a similar pattern to those patented in the USA with a wooded 'T' handle. Advertising emphasised the mowers ability to collect cuttings in a box, or distribute them, also for easy transportation the mower could simply be flipped over and rolled out of gear.

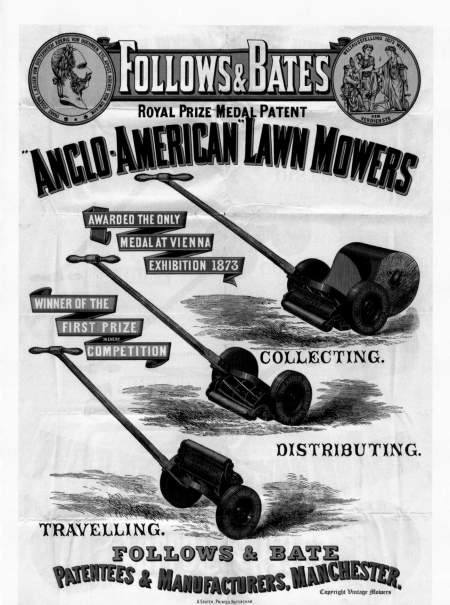

MOWERS IMPACT ON SPORT

Whilst our grass-based sports were played long before the lawn mower appeared, the mower has had considerable impact on their development.

Bowls, croquet, tennis, golf, cricket, football, rugby and many more have seen their playing surfaces change and develop as the improvements and efficiency of the lawn mower grew.

By the early 17th Century the game of bowls was already secure in the affections of all levels of society with aristocrats commissioning bowling greens in their grounds. Although Richard II banned it claiming that it was distracting people from archery, the prohibition was renewed by Henry IV & Edward IV, and re-imposed by Henry VIII.

Did you Know?

From 1600, grass was the preferred surface to play on, although it is often said that Drake may well have played his immortal game on an expanse of camomile.

Whilst lawn mowers were no doubt used on sports turf from their invention in the 1830s, advertising and the printing of testimonials extolling their virtues was perhaps aimed more at the large country estates owned by royalty, dukes and lords.

Mowers on the 10th Green Carnoustie Golf Club

By the 1870s Ransomes were recommending their recently introduced 'Automaton' as being good for the mowing of croquet lawns, a smaller manufacturer B. Hirst & Sons of Halifax in an 1870 advertisement stated that

PATENT
"CHAIN-TENNIS" LAWN MOWER.

their machine was acknowledged to be the best for keeping Bowling Greens and Croquet Grounds in order.

In 1874, Follows & Bate an established mower manufacturer based in Manchester, advertised their croquet mower stating that it was remarkably easy to work and especially designed for ladies' use! A few years later in 1882 they introduced a new mower appropriately named the 'Tennis' initially only available in six, seven and eight inch cutting widths.

It became clear there was a growing market required to support the sporting industry no doubt stimulated by the mower's invention and many companies began designing specific mowers for the purpose, whether it be a Ransome's Bowling Green or,

Shanks Lynx Mowers - Centre Court Wimbledon

Putting Green Mower, a Shanks Golf Lynx mower, and their 'Drake' models, designed for bowling greens.

Motor mowers continued in helping sporting venues to grow and become more efficient with Atco publishing a testimonial from The All England Tennis Club praising their Atco Standard which was said to deal with an area of turf of 1,000 square yards in twenty minutes with a fuel cost of only 1¼d.

Dennis motor mowers introduced in 1921 became the norm for the upkeep of fine cricket wickets.

Horse drawn mowers even after the invention of the motor mower were still the first choice for large expanses of grass, golf fairways etc, although they would be superseded by the gang mower which is described in more detail later in the book.

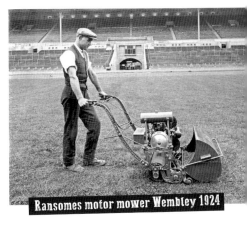

Ransomes motor mower Wembley 1924

The above photograph taken in the mid 1920's shows two 42in Thomas Green & Son's motor mowers at Lancing College, West Sussex, used for the upkeep of their football, rugby and cricket pitches. One of the mowers is believed to have survived and is held in the Lumsden mower collection at Picton Castle , South Wales, having been restored by the late David Lumsden.

ON THE EDGE

The final touch after mowing your lawn is often to trim the edges to give a definitive sharp edge; nowadays this is often achieved with a strimmer or long handled edging shears.

As early as 1825 (five years before the lawn mower was invented) an edging iron was invented by Charles McIntosh, gardener to Sir Thomas Baring, Bart. M.P. In a letter to the *Gardeners Magazine* he says *'I now send you a drawing of the edging iron, of course some allowance will be made to the first trial, as men are apt to be prejudiced against new tools; but I can pledge my word, that I will myself (notwithstanding my infirmities) cut as much in one day with this instrument as I could in four, or I may say in five days with the instrument in general use.'*

Edging Iron

A variety of edge cutting equipment appeared during the 1870s and 1880s from various manufacturers including Thomas Green & Son and Ransomes, Sims & Jefferies.

FIG. 160.—RIDGWAY'S LAWN-EDGE CUTTER.

1882

Townsend's Verge Cutter.

"We can strongly recommend Mr. TOWNSEND's Machine, as being effective in execution and a marvellous saving of labour, which in these days is an important consideration."—*Journal of Horticulture.*

Wimbish Ironworks, Saffron Walden. **1874**

ADIE'S PATENT LAWN EDGER.

THIS Machine constantly employed will pay itself in two days. Dr. Hogg, in the *Journal of Horticulture*, says—"This Edge Clipper we have tried, and know not which to admire most—its simplicity or efficiency." Mr. Moore, in the *Florist*—"This new machine does its work rapidly and admirably, the grass being cut with precision," and he further adds, "the use of it will, we have no doubt, become general." Prices 25*s.* and 30*s.*
LAWN EDGER CO., 15, Pall Mall, London, W. **1881**

GREEN'S PATENT GRASS EDGE CLIPPER.

1883

SIZE and PRICE.

Wide.   Diam.
8 inch .. 7 inch .. £1 16s.
Packing Case, 2s.

Specially designed to meet a want that has long been felt in cutting the overhanging grass on the edges of walks, borders, flower-beds, &c., and do away with the tedious operation of cutting with shears.

ADIES' LAWN EDGER.

PATENT.

Awarded BANKSIAN MEDAL of Royal Horticultural Society.

1880

GREEN'S PATENT GRASS EDGE CLIPPER

Specially Designed for

Cutting the Overhanging Grass on the Edges of Walks, Borders, Flower Beds, &c.

It is simple in construction, is easily worked, and reduces labour immensely.

1878

Size—8 inches wide, with roller 9 inches diameter, £1 10s. 1½
☞ Delivered, Carriage Free, at all the principal Railway Stations and Shipping Ports in England, Ireland, and Scotland.

THOMAS GREEN & SON,
SMITHFIELD IRONWORKS, LEEDS;
And 54 and 55, BLACKFRIARS ROAD, LONDON, S.E.

RANSOMES' NEW PATENT EDGE CUTTER.

The only really practical Edge Cutter yet brought out.

Any Gardener can use it with ease.

THIS Machine is a light, simple, strong, and useful substitute for Hand Shears, for trimming the edges of Lawns. It can be guided with the greatest ease either in a straight line or round the edges of flower beds, &c. The depth of cut can be varied to suit the depth of the edge by simply tilting the conical roller more or less. The knives can be adjusted to the fixed blade as they wear, and can be readily removed for sharpening. After a little practice with this Machine, a man can trim the edges as fast as he can walk, thus doing from five to ten times as much work as with the ordinary shears.

Price ...   35s.
Carriage Paid to any Railway Station.
RANSOMES, SIMS & JEFFERIES (Limited). Ipswich.

1885

Green's early edgers from the late 1870s to early 1890s were based on the double handle mower design with a heavy roller 7in wide, those produced from 1883-1894 sharing parts with their Multum-in-Parvo and possibly the Monarch mower. In 1895 a new improved design with wooden 'T' handle was introduced known as 'The Handy'; this was still available in Greens 1936 catalogue with just minor improvements.

Ransomes, Sims & Jefferies also commenced with a doubled handled edger around 1885. This had a curious conical roller

which was in fact the depth of cut adjustment, made by simply tilting the handles to one side on the conical roller! For a short period at the turn of the century a traditional round roller was used with a chain drive, and then followed by a gear drive version which continued into the 1940s.

A short lived edger was produced by Suffolk Iron Foundry and patented in 1939. The inventor was John Henry Wyndham of Tunbridge Wells, Sussex, and his design used a miniature cutting cylinder turned at right angles to the direction of travel, enabling it to trim the overhanging blades of grass.

Thomas Green's Edger 1894

It would appear this edger was only produced for a very limited period, no doubt dropped from production during and following World War II.

The Suffolk idea would appear to have been copied in a patent granted to Patrick Robert Reid of Uckfield, Sussex, twenty years later in 1959, where no reference to the earlier patent was made.

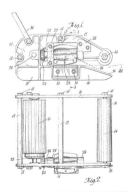

Wyndham Patent 1939

This patent turned up in my research shortly before I was due to give a talk to the Uckfield Horticultural Society and in view of the local connection I made reference to Patrick Reid in the talk and asked if anyone knew

Colditz Castle

of family connections locally. I was faced with a clear response of 'Colditz' from several members of the audience, had I touched a nerve or done something wrong? It was then revealed to me that the said Patrick Reid was in fact the famous army officer who escaped from Colditz (the German high security prisoner-of-war camp for officers who were considered to be a high security or escape risk).

Patrick Reid

Pat Reid was escape officer from 1940–1942 helping many colleagues with escape attempts; he did spend a month in solitary confinement with only bread and water when an attempted escape was discovered. He finally escaped on 15 October 1942 arriving in Switzerland on the 20 October; he remained there until 1947 eventually settling in Sussex

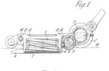

Fig. 1

Fig. 2

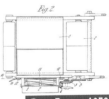

Reid Patent 1959

John Mills

when he left the army. Pat Reid is portrayed in the film *The Colditz Story* by the famous actor John Mills.

Is it just a coincidence that Pat Reid lived a mere 16 miles from where John Wyndham was living when he applied for his patent?

STEAM POWER

A steam engine is a machine that converts the heat energy of steam into mechanical energy. Water is heated by burning of wood, coal, paraffin or petrol, the steam is passed into a cylinder, where it then pushes a piston back and forth. It is with this piston movement that the engine can do mechanical work. The steam engine was the major power source of the Industrial Revolution in Europe in the eighteenth and nineteenth centuries. It dominated industry and transportation for 150 years.

The first steam-powered machine was built in 1698 by engineer Thomas Savery (1650–1715). His invention, designed to pump water out of coal mines, was known as the Miner's Friend.

Thomas Newcomen (1663–1729) made further developments, followed by James Watt (1736–1819) speeding up the Industrial Revolution both in England and the rest of the world with his contribution.

Steam was successfully adapted to power boats in 1802, railways in 1829 and road transport vehicles from the 1850s.

John Claudius Loudon (1783 –1843) botanist, garden designer and author of *The Gardeners Magazine*, reviewed Budding's invention of the lawn mower in his magazine in 1831, stating: '*I have little doubt that it will soon be modified to be worked by donkeys, ponies or by small steam engines; but whether it be or not, it promises to be one of the greatest boons that science has conferred on the working gardener in our time*'.

Green's Steam Mower

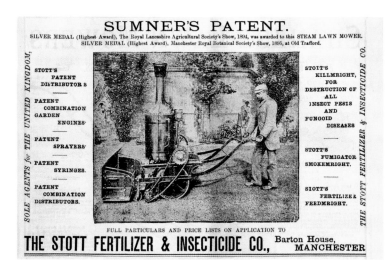

It was sixty years later that James Sumner of Leyland, Lancashire, built the first steam lawn mower in 1893, being produced during the following few years by the Lancashire Steam Motor Company, many years later to be part of the British Leyland Group.

Other mower manufacturers were keen to produce their own steam powered mowers with Thomas Green & Son sending representatives to inspect the Sumner mower and by 1903 they were producing their

Did you Know?

Sumner was a talented engineer, who from the 1870s produced many different vehicles running on steam, wagons, cars and even a tricycle on which he was fined for speeding at 8mph!

own in 30, 36 and 42-in cutting widths. Shanks of Arbroath after two years of development introduced a 36in model in 1900 but the experimentation of using the internal combustion petrol engine was well underway and the steam mower was short lived.

There was much debate in the pages of the *Gardeners Chronicle* in 1896 for and against the 'Steam Mower' as shown below.

Sumner Steam Mower, on display at Museum of English Rural Life, Reading, Berks.

Now gardeners, allow me to ask you a pertinent question. Are you wishful to retain your standing as garden craftsmen, or are you going to reduce yourselves to the position of tenders of steam-engines?

Taking an unbiased view of the different motive powers, I have no hesitation in saying that where there are large lawns to be kept in order, the use of steam as the motive power is most advantageous,

In the case of private gentleman's pleasure-grounds the lawn mower, whether propelled by manual or horse-power, is a most efficient implement. It is also more in harmony with rural surroundings than one propelled by steam-power would be. Should it be a question of economy, the price of a steam lawn-mower would pay a couple of good labourers for a year. Mr. Knapp speaks feelingly of the toiling labourer, working an 18-inch machine, but such work in the open air would hurt no man. Of course, the iron-horse on the iron-road is an important factor in our daily life, but we still want a few quiet corners where men have time to think and grow young again; and we do not want all the land covered with iron-roads, steam-engines, and cinder-heaps. *R. M., Newbury.*

I cannot bring myself to see that the introduction of steam as the motive power for lawn-mowers in any way lowers the position of a gardener as a craftsman, because very little craft is required to steer a machine, whether it be moved by manual, animal, or steam-power.

I do not think a case has been made out for its application to lawn-mowers, and I hope the day is distant when steam lawn-mowers will be puffing and snorting in the sylvan retreats of rural England; for if the quiet at present enjoyed in the surroundings of country mansions be broken up by steam lawn-mowers and other complicated machines.

THE STEAM LAWN-MOWER.—The fact that steam is being put to a good use in propelling the heavy, lumbering mowing-machine of cast-iron and steel is getting up heat among those who suppose that much-valued quiet of the garden is going to be disturbed for ever and aye by the snorting, smoking, abominable thing working all day long, and always just in front of the drawing-room windows.

I know from experience that there is not much poetry in the movement of a 22-inch lawn-mower when you are responsible for half the motive-power on a hot day, but it is good exercise for a strong, healthy man. *R. M., Newbury.*

Gardeners Chronicle 1896

PEACH'S PETROL POWER

In 1829, whilst Edwin Budding was busy inventing the lawn mower down in the Stroud valley, another engineering genius was busy 300 miles away in Newcastle-upon-Tyne putting the final touches to his invention, the most famous steam locomotive of all time 'The Rocket'; he was, of course, George Stephenson.

Move on just over twenty years to 1852 and William John Peach was born to parents George Peach and Ann Stephenson, at the age of twenty-six in 1878 he changed his name by deed poll to Stephenson-Peach.

In later years various articles have been published stating that William John Peach was the grandson of the renowned George Stephenson; the research dispels this as a myth, primarily because there is no record of George having any grandchildren.

However, William's, Grandfather on his mother's side, John Stephenson (1794-1848) was involved with the development of the Sheffield and Rotherham railway, in particular earthwork construction and deep cuttings.

Stephenson Peach Roller Mower 1894

Whilst John does not appear to have any family connection with George Stephenson either, his brother-in-law was Isaac Dodds (1801-1882) who became George Stephenson's first apprentice and in fact accompanied Stephenson to Killingworth pit for the first test of the miner's safety lamp that he had invented. Like Budding, Isaac Dodds was scarcely appreciated during his lifetime but was a talented engineer being instrumental in the development of engine's coupling rods, turntables and the sprung loaded buffers.

William Peach, a young talented engineer, would have been well aware of George Stephenson's fame and perhaps a bit of poetic licence was adopted in his change of name.

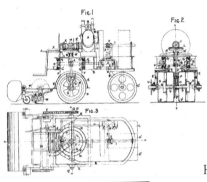

Stephenson Peach Patent 1895

Either way it certainly did no harm to William John Stephenson-Peach's career and his contribution to lawn mower development, having designed and patented the first petrol-engined lawn mower in 1895.

He created practical engineering workshops at public schools including Repton, Cheltenham and Malvern College. During the early 1890s he designed steam, electric and petrol driven lawn mowers and rollers. In 1895 he successfully patented one of the first petrol driven motor mowers which was combined with a heavy roller.

In February 1896, a report was published by a Mr G. Don who had met with Stephenson-Peach for a demonstration of his mowers.

Mr Don was impressed with the petrol driven machine stating, *'The machine is managed with the greatest of ease, a person seated on the front part of it, with a lever in each hand, starts guides and stops it at will and in perfect certainty. The facility with which it is steered is perfectly marvellous; to show what it could do in this way, it was made, at one part of the trial, to turn round and round on one spot not much greater in diameter than its length. There is no smoke, and the noise made by this machine, which cuts and rolls a width of 3½ feet, is no greater than would be caused by a horse-machine of equal width.'*

In 1895 he also patented an electric roller, a well circulated and published photograph of the roller shown has to date stated that Thomas Green (Thomas Green & Son) is the well-dressed gentleman in the driver's seat, this has been in doubt since Green died in 1892. Research by the author reveals it is in fact more likely to be Stepehnson Peach's father-in-law a Mr Symington.

Stephenson Peach Electric Roller being operated by his Father-in-law, Mr Symington

Whilst Stephenson-Peach's machine does not appear to have gone into production, it is understood in 1897 Coldwell Lawn mower Co. of New York produced its first motor lawn mower which was on very similar lines.

Stephenson-Peach who had already designed and made motor cycles and three wheelers in his workshops at Repton, attracted the attention of Henry Morgan and was asked to help construct what was to become the very first of the famous Morgan three wheeler cars.

CONTINUED DEVELOPMENT
1900 TO 1960

RANSOMES, SIMS & JEFFERIES GET MOTORISED

James Edward Ransome of Ransomes, Sims & Jefferies patented designs for a 42in commercial ride on motor mower and a smaller pedestrian mower, probably 24in, in 1902 and the company are credited with having put in production the first commercial motor mowers. Their first customer was a Mr Prescott-Westcar of Herne Bay who purchased a 42in machine powered by a 6hp 4-stroke water cooled engine made in Germany by Simms.

By coincidence the directors of Thomas Green & Son met in May 1902 at William Penrose Green's house to inspect a new motor mower powered by a German 'Fafnir' engine.

It is clear that there may well have been some collaboration between the mower producers, with the three main manufacturers Shanks, Greens & Ransomes meeting every year to agree terms of business for the following year, adopting as far as possible uniform catalogue prices.

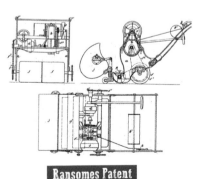

Ransomes Patent

By 1907 Ransomes had supplied around 150 motor mowers including the supply of two 30in motor mowers to His Majesty King Edward VII with numerous others being purchased by the nobility.

The 7th Earl Fitzwilliam acquired a 30in motor mower from Ransomes for use at his home in South Yorkshire, Wentworth Woodhouse now Grade 1 listed and considered to be the largest private residence in the United Kingdom. The house has more than 300 rooms, although the precise number is unclear, with 250,000 square feet (23,000 m^2) of floorspace (124,600 square feet (11,580 m^2) of living area). It covers an area of more than 2.5 acres and is surrounded by a 180-acre park, and an estate of 15,000 acres.

Ransomes 42 with Simms engine

Wentworth Woodhouse

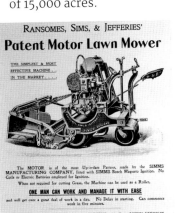

RANSOMES, SIMS, & JEFFERIES'

Patent Motor Lawn Mower

THE SIMPLEST & MOST
EFFECTIVE MACHINE ..
IN THE MARKET ..

The **MOTOR** is of the most Up-to-date Pattern, made by the SIMMS MANUFACTURING COMPANY, fitted with SIMMS Bosch Magneto Ignition. No Coils or Electric Batteries employed for Ignition.

When not required for cutting Grass, the Machine can be used as a Roller.

ONE MAN CAN WORK AND MANAGE IT WITH EASE

and will get over a great deal of work in a day.   No Delay in starting.   Can commence work in five minutes.

Machines have been in constant use during last Season at KEW GARDENS, and by Messrs. CADBURY, BIRMINGHAM, and S. PRESCOTT WESTCAR, Esq., HERNE, with the greatest success. *Every facility afforded for trial.*

MADE IN THREE SIZES, one as shown for man to ride, and the other two arranged for man to walk behind.

PRICES ON APPLICATION.

RANSOMES, SIMS & JEFFERIES, Ltd., Ipswich.

The Earl was certainly pleased with his mower with several photographs having been taken in around 1904 with his head gardener, a Mr J. Hughes in attendance. The following testimonial appeared in Ransome's 1907 catalogue...

Ransomes Ad 1903

The Right Hon. Earl Fitzwilliam, Wentworth-Woodhouse, Rotherham (Mr. J. Hughes Head Gardener)

The 30in Motor Lawn Mower you supplied for use in these Gardens has been running with few intermissions, the whole of the present season, and has done its work admirably. It is a great advance on horse-drawn mowers, inasmuch as it gets through the same work in a much less time, does not disfigure the turf as horses' feet do when wet and soft, and is less expense to work. It is easily managed by the garden men, and has so far given every satisfaction.

Ransome's Patent Motor Mower Wentworth Woodhouse

THE GANGS MOVE IN

As stated earlier, the traditional horse-drawn mower still reigned on the large expanses of grass, parks, sports fields and golf fairways until the 1920s when the horse-drawn gang mowers became available. The gang mower was basically a frame that linked together a number of cylinder mower cutting units, cutting in tandem thereby being able to cover a much wider area with each pass.

Charles Worthington applied to patent the idea in 1913 and later the machines were produced by The Worthington Mower Co. of Oklahoma, USA.

Whilst initially pulled by a horse, over time a vast variety of specially modified vehicles were used to pull the gang mowers, tractors, trucks and adapted cars, some based on the Austin Seven and Ford Popular.

Ransomes of Ipswich were granted a licence to manufacture gang mowers under the Worthington

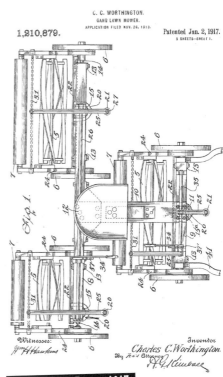

C. C. WORTHINGTON.
GANG LAWN MOWER.
APPLICATION FILED NOV. 26, 1913.

1,210,879.

Patented Jan. 2, 1917.
5 SHEETS—SHEET 1.

Fig. 1.

Witnesses:
Wm H Hawkins

Inventor
Charles C. Worthington.
By his Attorney
A. G. Kimball

Worthington Patent 1917

The
LLOYD-BEDFORD TRACTOR
embodies improvements which place it
in a class by itself
**FOR ALL-ROUND SERVICE AND MINIMUM COST
OF UPKEEP.**

It has ample power, remarkable flexibility and maximum
efficiency with complete absence of vibration.

A WIDE RANGE OF GEARS
enables the most suitable speeds to be maintained for all
varieties of work, from 2 m.p.h. drawing gang mowers over
soft turf to 30 m.p.h travelling on a hard road. The twin
pneumatic-tyred rear wheels are suitable for both purposes.
A Practical Patent Attachment for Cutting Bents can
be fitted.

LLOYDS & CO. (LETCHWORTH) LTD
PENNSYLVANIA WORKS ·· LETCHWORTH

patent in 1921 and within ten years they were producing, triple, quintuple, septuple units with overall cutting widths of 7ft, 11ft 6in and 16ft respectively. They added a Nonuple gang mower with nine units cutting a width of 20ft in 1939.

Other manufacturers soon followed; both Thomas Green & Son and Shanks of Arbroath added gang mowers to their catalogues in 1923. Lloyds of Letchworth specialised in the production of gang mowers with the accompanying picture showing an eleven unit gang with a cutting width of 26ft at work on Middlesex aerodrome.

In 1956 the Gang mower with one man could cut 11 acres an hour.

FACT

Lloyd's 11 Gang 26ft.

Ransomes Quintuple Pattison Tractor

Ransomes Overgreen in use

In 1937 Ransomes introduced their two wheeled Overgreen, specifically designed for golf greens, again under licence from the Worthington Mower Company. The Overgreen was a self-propelled walk behind tractor unit with a single cylinder 348cc four stroke engine; it was equipped with one forward and two trailing 16in Ransome's Certes fine turf mower cylinders giving a total cutting width of 3ft 8in.

Advertising in 1940 stated that the Overgreen outfit would enable one man to cut eighteen golf greens in one day thereby releasing two of three men for other important or urgent work.

18 greens cut —in one day—

ONE man with an Overgreen can cut 18 average putting greens in less than a day. Two or three men can thus be released for other important work on the course.

Three 16 in. Ransomes Mowers are operated at one time, universally connected to allow them to float smoothly and independently over undulating greens. The power unit consists of a specially designed 2-wheel tractor with low pressure balloon tyres.

Ransomes

OVERGREEN MOWER

For Export
to all countries of the World

Illustrated literature showing this, and our complete range of hand and motor lawn mowers, gang mowers, etc., will be sent on application.

RANSOMES, SIMS & JEFFERIES, LTD.
IPSWICH _____ ENGLAND

During the war years lawn mower production by the major manufacturers was reduced considerably or ceased completely.

In August 1916, a leading mower manufacturer Ransomes, Sims & Jefferies were asked to make aircraft for the Royal Flying Corps with another Ipswich company, Frederick Tibbenham, and a new factory was built off Cavendish Street.

Did you Know

During the war Ransomes produced munitions including: 790 aeroplanes, 650 airship sheds and aircraft hangars, 3,700 aircraft bombs, 440,00 shell cases, 300,000 shell and fuse components, 5,000 general service wagons and 10,000 bombs for Stokes guns.

Thomas Green & Sons, like others, experienced a shortage of materials and by 1918 they had a back order for 10,143 mowers that they could not fulfil. One of the solutions was to produce more gear driven machines as these could be made from in-house castings when chain steel was not available.

During the war in many situations life had to continue with grass to be mown, an excellent example being a girls' school in Surrey where the teachers and girls formed the 'Field Volunteers' and took over the ground keeping work whilst the men were at war. The image attached depicts the Field Volunteers in 1918 armed with a selection of equipment. For the mower enthusiast from right to left, Greens New Century Sidewheel mower, Thomas

Green's Pony Mower, White Line Marker, Heavy Roller and a Thomas Green's Silens Messor mower, which incidentally in the catalogue of that era would have suggested operation by a 'Man & Boy' not a 'Woman & Girl!'.

World War I's 1918 Field Volunteers

World War I's 71st RFA

Another mower-related image from World War I shows a small group of soldiers from the 71st Royal Field Artillery tending the gardens within their barracks, again with a Green's New Century push mower. It is interesting to note that the 71st RFA went on to see active service in France and the war diaries indicate they played a valuable part in the war.

PUSH WITH POWER

During and after World War I social changes meant that fewer people were being employed to work on gardens and estates. Owners of large estates and country houses looked for labour-saving alternatives to existing manual mowers that often required two people, one pulling, the other pushing.

Ransome's Mower Pusher

Two companies introduced Mower Pushers in the early 1920s; William Edgecumbe Rendle obtained a patent in 1914 for a motor-driven machine to propel lawn mowers and other slow moving machines, the first being produced in 1920. Shortly afterwards Ransomes, Sims & Jefferies produced a similar machine, the MP Mower Pusher marketed by MP Co. in London's Oxford Street.

The pusher could be connected to most makes of roller mower with special brackets, as depicted in the accompanying image, this shows an MP pusher connected to a Ransome's Patent Chain mower.

Mower pushers only enjoyed brief popularity, probably due to the reduction in the price of motor mowers in the mid-1920s, when mass production reduced their cost considerably.

MOTOR MOWER FOR THE MASSES

Atco 1921 diagram

In 1921 Charles H. Pugh Ltd introduced their first mower, a petrol engine powered mower which later became known as the Atco Standard and indeed the first mass produced motor mower. By 1926 they had sold 16,000 in a variety of sizes with the 14in model priced at 30 guineas (£31.50).

The Pugh family had been involved in engineering since the mid 1800s with Charles Henry Pugh inventing the seamless bicycle wheel rim in 1894, during which period he owned the Whitworth Cycle Co. later to become Rudge-Whitworth.

In the early 1900s the company was best known as a successful manufacturer of small 'repetition' components. These were

The **ATCO** MOTOR MOWER

a money saving investment.

Grass Cutting Time is round again.

Start the season aright by having a motor mower —the Motor Mower—the "ATCO" Motor Mower.

One man with an "ATCO" equals three men with push machines. Any ordinary workman can operate it perfectly —even a child can do so—for the "ATCO" Motor Mower is as easy to understand as a push mower. Besides costing surprisingly little to buy and to operate, the use of an "ATCO" improves your turf and enhances the beauty and value of your property.

Over 1,000 "ATCOS" were in use last year.

As a Guarantee of the excellence of our materials and workmanship, we undertake to supply replacements free of charge within one year from the date of our Agent's sale of any parts failing through faulty material or defective workmanship.

PRICE: £75 Carr. Paid.

YOU TAKE NO RISKS WITH AN "ATCO" We will give you a free demonstration on your own grass and maintain regular attention from Service Depots throughout the Country. Write to-day for FREE Booklet, to

CHARLES H. PUGH, Ltd., Whitworth Works, 10 Tilton Road, BIRMINGHAM. *22 in. Cutter. Nine Blades.*

WHAT SATISFIED USERS SAY. *"Very pleased with the Atco Mowing Machine. In three hours yesterday it cut as much grass as it usually takes two men a day-and-a-quarter to cut."* Mr John Holder, Berks.

Atco Standard Ad 1922

Atco Ad 1929

The **LAWN MARK** *of* **DISTINCTION**

THE lawns of many of the stately homes of England, the leading Sports Grounds and Golf Clubs, are maintained in their perfect condition by the famous ATCO Motor Mower.

But ATCO means more than the perfect lawn. By its simple efficiency it introduces a definite economy of labour, mowing time and costs.

Ask for the ATCO Catalogue, which describes the seven models —one of which is ideal for your requirements—or let us arrange a Free Demonstration on your own lawn.

ATCO ALL BRITISH MOTOR MOWERS

CHARLES H. PUGH, LTD., 8, TILTON ROAD, BIRMINGHAM.

---

used extensively during World War I, on many items, including armaments. In 1911 they patented the 'Senspray' carburettor, a major development in fuel delivery to small internal combustion engines; this was used extensively on early motorcycles, particularly with Villiers engines.

During the same period Pugh's also produced the Alfred Appleby Cycle Chain advertised as 'The Best Cycle Chain in the World'. The decision to start making mowers in 1921 was apparently taken after the pony that pulled the mower cutting the grass around the factory died; it was suggested as the company already made various components, chain,

engine parts etc, surely they could design and build their own motor mower, rather than replacing the pony.

The Atco mower business flourished in the following forty years during which time they became a household name and a major supplier to the domestic market.

Sales were backed by a vast servicing department supported by a fleet of vans and also Rudge-Whitworth motorcycles with specially adapted sidecar outfits to carry a motor mower. Servicing wasn't just a quick sharpen up of the blades, but also included a thorough clean, mechanical service and repaint with new transfers.

Atco Std. with two models

Rudge Whitworth Motorcycle Service

Whilst the company experienced many takeovers and mergers during the last ninety-seven years, mowers with the Atco name are still being produced today.

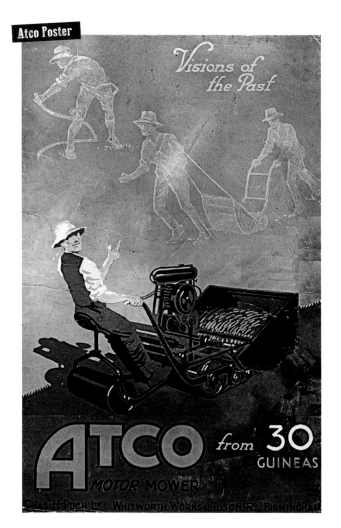

Did you Know?

An Atco Standard was used to prepare the pitch for the 1926 FA Cup Final at Wembley; Bolton Wanderers beat Manchester City 1-0 in front of 91,447 people.

BETWEEN THE WARS

During World War I many new engineering businesses had sprung up or adapted to make munitions and machinery for the war effort, with the war ended they now looked to manufacture other items.

The JP Super is another design classic and one of the most famous of all lawn mowers. It was introduced in the 1920s and continued in production until well into the 1950s.

The mower was made by a company formed by two engineers named Jerram and Pearson (hence the name) in Leicester, England, in the years immediately after World War I. The company prided itself on the quality of its design and engineering and JP mowers were known as the 'Rolls Royce of Lawn mowers'. In fact, the two companies did have links through other engineering interests.

JP Poster

When the JP Super became available in 1920 it was an instant hit. The extensive use of aluminium and alloys in the JP Super was a revelation in its day. The company's brochures proclaimed the virtues of the nickel steel, steel drop forgings, pressings and aluminium alloys they used that were to 'Air Board Specifications'. In the 1930s, they claimed that 'metals employed in making JP Super lawn mowers are largely the same as used in British Schneider Cup machines', this being the series of races for aircraft held in the 1920s and 1930s.

The mower also featured a number of design innovations. For example, it incorporated high precision components, especially in the epicyclic gearing, ball bearings and roller chain that transmitted power from the rear roller to the cutting cylinder inside a heavy aluminium casing. This was at a time when many rival products were still utilising cast iron sprockets and block chain, a concept dating back to the 1850s.

Another innovation was the quick release mechanism of the cutting cylinder. By simply undoing the large nut on the side of the mower, the cylinder could be released without removing the chain or sprockets. This simplified maintenance and repair. The idea has been copied many times by other manufacturers, but never bettered.

Singer Mower

All of this engineering made the JP Super a high quality machine, but also very heavy and relatively expensive. The company itself admitted that the 'cost of these materials alone is, in many cases, more than the entire cost of cheaper and so-called competitive machines'.

Nevertheless, the JP Super was a success. It was available in 12in and 16in cutting widths although the smaller model was by far the most popular.

JP Super push lawn mowers are still in regular use and are the pride and joy of many keen gardeners, although production actually ceased in the late 1950s.

Many other established companies tried their hand at producing lawn mowers in the 1920s and 30s; Dennis Brothers of Guildford is one example but they are perhaps more famous for their commercial vehicles including fire engines. Royal Enfield was another deviating from their main business of making firearms and motorcycles to produce their own lawn mowers.

Another company Singer & Co. of Coventry also produced a motor mower in the late 1920s although to date only one is known to have been made, perhaps a prototype, now held in the Museum of Gardening in Sussex.

Dennis Mower

ELECTRIFICATION

Stephenson–Peach had dabbled with large electrical rollers and mowers in the 1890s with only a few being made.

Electric Coldwell Mower

It would appear the first production mains electric lawn mower was designed by Coldwell in the USA and launched after two years of development, patent applied for in 1925, the machine continued in production until 1932. The mower was driven by a specially designed General Electric Universal motor of ½ HP running on 110 volts AC or DC, being the normal house lighting circuits at the time. It came with 150 feet of cable and was said to consume the same amount of power as an electric iron.

Ransomes, Sims & Jefferies were soon to follow having a patent granted in 1927 for their version of a mains electric lawn mower the 'Electra', this was followed in later years by the 'Lawnic' and' Bowlic' specifically designed for the fine turf of tennis courts and bowling greens.

The 1940s to 1960s saw numerous companies introducing electricity to the lawn mowing public; cleaner and less fuss than petrol, with reduced effort compared with

The "ELECTRA" 14-in. MODEL

16-in. and 20-in. MODEL

Ransome's Electric

the push mower.

A rather futuristic electric mower, the Ladybird, appeared in the late 1940s, the manufacturers emphasising its lightness at only 15lbs (6.8kg) and as easy to use as a vacuum cleaner. The company also provided electric conversion kits for popular push mowers.

Ransome's Electric - Thornbury Bowling Club, Melbourne Australia

Ladybird Ads

THE 'Ladybird' MOWER by A.M.I

REGD. TRADE MARK

PATENTS PENDING

A MOVE DOWN UNDER

In the aftermath of World War I, Australia looked to become more self-sufficient in manufacturing both by their own companies expanding and looking for overseas companies to bring their own production to the country.

Qualcast French Ad

One of the firms targeted by the then Commonwealth Minister for Customs and Trade Mr J.E. Fenton on a trip to the UK in 1930 was Qualcast Ltd of Derby who had been making lawn mowers since the 1920s. Qualcast's main mower production was focused on the low priced side wheel mowers, this being the type favoured by the Australian government as suitable for the masses rather than the more expensive roller mowers being produced by other UK manufacturers. There did appear to be a carrot and stick scenario with possible heavy tariffs to be imposed on future mower imports; Qualcast agreed the terms and commenced plans to set up a factory to produce their mowers in Australia.

Qualcast Display

The whole new manufacturing plant (a duplicate of the Derby works) was carried on board the merchant vessel Mahout with the capacity to produce 500 lawn mowers per week and employ 150 local people. The factory was opened by acting Prime Minister James Fenton on 6 October 1930 and occupied an acre and a half at Footscray, Victoria.

Did you Know

By 1934 production was increasing steadily with the Australian-produced mowers selling at two-thirds of the cost of an imported mower.

Qualcast certainly gained worldwide appeal with their excellent marketing selling many mowers throughout Europe, in particular the French market.

A new mower manufacturer entered the market in the early 1930s having patented their first lawn mower towards the end of the previous decade, H.C. Webb & Co. of Birmingham. I have had the pleasure of meeting the descendants of Frank Walter Taylor

Frank Taylor

(1895-1963) who was the driving force behind the development of Webb mowers with over 140 patents to his name, working for them for over forty years.

Webb Ad 1933

H.C. Webb was the first manufacturer to mass produce lawn mowers with pressed steel side frames rather than cast iron, which had been the norm for almost 100 years, this was a major advancement for both push and motorised mowers.

Webb's early hand mowers all had names starting with a 'W' – Witton, Witch, Wasp, Whippet and Windsor.

A popular Webb children's mower was produced in 1957 and appeared at the British Industries Fair in Birmingham, based on one of the standard models, with four cutting

Webb Miniature

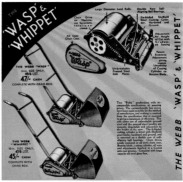

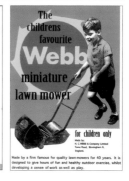

blades and grass box it actually cut the grass. As a safety measure it had a fixed belt drive so that when the machine stopped moving, the blades stopped instantly. Should anything get trapped in the blades the belt slips and they stop to hopefully avoid injury. The handles could be set at heights from 21in to 27in and the height of cut could be varied as on a real machine.

The toy mower certainly had royal approval, with Prince Andrew and Prince Edward seen playing with one in 1966 whilst celebrating Prince Edward's 2nd birthday in the grounds of Buckingham Palace. Photograph taken by Lisa Sheridan (1893-1966) Royal photographer for over 30 years.

Did you know?

Webb's early hand mowers all had names starting with a 'W' Witton, Witch, Wasp, Whippet, and Windsor.

Evie with Toy Mower

THE ROTARY

The 1930s became a time of development and innovation in the engineering world with an army of inventors patenting new tools and machinery. The lawn mower was to undergo a major change in how the mower blade cut the grass, since Budding's invention lawn mowers had utilised a cutting reel working against a fixed knife bed to achieve the cut, with a few using a reciprocating blade similar to hay and straw mowers.

The breakthrough came in 1932 as a result of an attempt by David Cockburn to make a rotary hedge trimmer which he combined with a domestic vacuum cleaner to cut his hedges and collect the trimmings. It was not a success but as he dragged his invention across a lawn with the motor still running he noticed that as well as cutting a strip of grass it also picked up the grass clippings.

Rotoscythe

Cockburn's patent was granted on 29 December 1932 with the first rotary mower being produced by Power Specialities of Slough in 1933; powered by a petrol 2 stroke engine.

THE
ROTOSCYTHE
VACUUM LAWN MOWER

Early advertising described the Rotoscythe as the vacuum lawn mower and stated that it *'Employs the principle of the scythe – agreed by all lawn culture experts to be the only entirely satisfactory method of cutting grass'* and continued to say that the mower could cut 2,000 square yards per hour.

Whilst it immediately gained popularity, initially rotary mowers did not produce the much admired stripes as did the traditional roller mower, this was rectified by some manufacturers in later years by fitting a rear roller to their rotary mowers.

FACT The blade of a rotary mower revolves 3,000 times per minute, with the tip of the blade cutting the grass at a speed of 200mph!

FINGER BAR

The finger bar or sickle bar mower has existed since 1845. Initially a rather crude form but later developed and used extensively in agriculture being drawn by a horse and later tractors. This type of reciprocating mower was and still is used mainly for agricultural crops. The cutting operation is very similar to modern day hedge trimmers with the cutting blade moving backwards and forwards over a fixed blade.

A move to horticulture happened in the early 1930s with pedestrian machines on a smaller scale being introduced by many different manufacturers.

One of the most famous was John Allen & Sons of Oxford who designed and patented the Allen Self-Propelled Motor Scythe in 1933–35, later to become known by many as the Oxford Allen Scythe. Whilst the primary purpose of the machine was to cut rough grass and scrub, in orchards and small holdings,

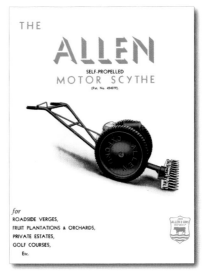

THE

ALLEN
SELF-PROPELLED
MOTOR SCYTHE
(Pat. No. 434079).

for
ROADSIDE VERGES,
FRUIT PLANTATIONS & ORCHARDS,
PRIVATE ESTATES,
GOLF COURSES,
Etc.

Allen Self-Propelled Motor Scythe

many attachments and extras soon appeared. Front mounted attachments included hay sweep, rotary brush, snow plough, yard scraper and cylinder mower; others were spraying, hedge trimming and even sheep shearing via a flexible shaft.

Lloyd's Autoscythe

When fitted with a trailing seat a trailer or set of gang mowers could be pulled.

Other traditional lawn mower manufacturers also entered the Autoscythe market in the 1930s, in particular Atco and Lloyds of Letchworth.

Atco Scythe

FACT

An advertising slogan for the Allen Motor Scythe declared 'Wherever a man can walk, an Allen will cut'.

A KICK START TO MOWER DESIGN

When Charles H. Pugh Ltd manufactured the first mass-produced motor mower in the 1920s the engine was started like many other makes by a crank handle, this often required a strong wrist action and, in some cases, led to minor injuries.

In 1935, Atco were granted a patent for a foot pressure starter which they suggested would be simpler, safer and more reliable. They were keen to state that in their trials every member of their female clerical staff had managed to start a machine with complete confidence. The new starting mechanism was fitted to their lightweight 14in model with advertisements suggesting that a push of the foot on the pedal made it as easy to start as a car. One wonders what type of car they were referring to as post 1920 most cars had electronic starters that did not require any physical pressure!

A push of the foot on the pedal and the engine starts. It is as easy as starting a car.

Atco Kick start Close-up

The so-called plunger starter only appears to have been used for around a year, after which a more traditional downward motion kick start pedal was introduced, this remained in use into the 1950s when it was replaced with a rope or cord pull start.

In the mid 1930s Atco were keen to inform their dealers of the lengths they went to in testing their new designs, in particular, the rotary track test as follows... '*The mower is run under its own power on a rough cinder track, without chain covers and with the grass box loaded, in all weathers, for the equivalent of three season's work.*

Twice during each lap the purposely unguarded chains dip into heaps of gritty dust and the rattle and shock transmitted to every point of the machine are infinitely more severe than actual working conditions of the worst sort.'

An image of the test track is shown below.

Atco Test bed

The company were also keen to ensure their customers took care of their new mower with the following instructions...

Housing your Atco

The 'stable' should be reasonably dry. The motor mower must not be covered with sacks of fertiliser, marl, lime, sterilised soil, leaf mould and so forth, when off duty. You may not think this happens but too often it does!

WORLD WAR II: 1939-1945

During World War II Ransomes, Sims & Jefferies with many of the male staff serving in the armed forces, employed more women in what had traditionally been male roles. Some were recruited from the Women's Industrial National Service scheme. The start of World War II saw the production of most lawn mowers stopped.

Production was centred on the manufacture of 17-pounder gun carriages, components for the Merlin engine used for Spitfire and Hurricane fighter planes, bomb trolleys, parts for crusader tanks and other items for the war effort.

Under the project name 'Farmer Deck' they fitted large ridging bodies and rollers on a frame at the front of Churchill tanks and in the lead up to the D-Day landings Ransomes made around 200 mine clearing devices for the tanks.

World War II - Churchill Tank

SOONER OR LATER YOU WILL BUY A HAYTER

In 1937, after serving an apprenticeship in the building trade, Douglas Hayter started his own building business in the village of Spellbrook in Hertfordshire. He erected workshops and sawmills, manufactured his own plant, and invented a top secret device for the War Office.

Douglas Hayter

World War II changed things and on learning that he was in a reserved occupation, Douglas Hayter decided to concentrate on buildings and repairing farm tractors and machinery.

When the war ended, he added the manufacture of cattle yard equipment to the company's activities. On discovering that some of his employees had no homes, he decided to build luxury caravans for them from old RAF mobile containers previously used on the *Queen Mary*.

In 1946, he decided to add some extra factory space at Spellbrook. He borrowed a cutter bar mower from a friend to clear a space for the new building but made slow progress with this work. He decided to see if he could design something better, with a second hand two-stroke engine plus a collection of bits and pieces that he had lying around; the first Hayter rotary lawn mower was created.

Hayter focused on the domestic market in 1957 with the introduction of the Hayterette, a hand-propelled 18in rotary mower, early models were red and silver and powered by a 98cc Villiers engine.

CONTINUED DEVELOPMENT 1900 TO 1960

The company developed the advertising slogan 'Sooner or later you will buy a Hayter' and soon afterwards designed the self-propelled roller drive Hayter mower range that achieved a striped finish to the lawn similar to that achieved by heavy roller mowers.

Here's Hayter's Latest!
The HAYTERETTE

CHELSEA SHOW
STAND 33

the Machine that is so much More than a Mower!

The 18-in. HAYTERETTE is powered by a Villiers 98 c.c. engine with a recoil starter. It has ample power for cutting its full width of 18 ins. under the toughest conditions—governed engine speed giving controlled power output with 12,000 cuts per minute!

Note these design features:

Large diameter rubber-tyred wheels with nylon bushes for minimum maintenance.

Wheels inset within frame enabling cutting to be made close to the base of obstacles.

Positive and easy cutting height adjustment by ¼-in. graduations in the range ¾ in. × 3¾ ins.

Dished steel bottom plate for minimum friction and maximum strength. Quickly replaceable cutter blades.

Alternative arrangement of handlebars—fixed or hinged. Cast aluminium alloy frame for lightness and durability.

CUTS IN ALL WEATHERS — CUTS ROUGH OR FINE — NO CHOKING. CHOPS AS IT CUTS — NO NEED FOR GRASS COLLECTION.

...and FULL 18" cut!

IDEAL FOR LARGE LAWN AREAS, VERGES, PADDOCKS, ORCHARDS, etc.

Please write for prices. Demonstrations arranged.

HAYTERS OF SPELLBROOK
8 Spellbrook Lane, Bishop's Stortford, Herts.

Hayterette Ad 1958

Did you know? **The Hayterette is still going strong today, fifty years on with the same design principle providing a neat finish to semi-formal lawn areas where collection of cuttings is not required.**

SABO RASENMÄHER

In 1932, five years before Douglas Hayter started his business in the UK, in Germany Heinrich Sanner and his partner Walter Born founded SABO (SA stood for Sanner, BO for Born = SABO) and their business initially focused on mineral oil products and technical innovations.

Heinrich Sanner

In 1946, Sanner like Hayter focused on horticulture; following the war many hedges had not been trimmed and Sanner considered how to do this time-consuming and tedious job easier and faster, the result being an electric hedge trimmer which was successfully manufactured by his company.

This was SABO's first point of contact with the so-called 'green market' and they quickly recognised the great opportunities that it offered. In 1954, Sanner developed his first motorised lawn mower focusing on robust and durable commercial machines.

With the growth of domestic gardening he looked to enter this market and in 1958 launched his first lawn mower with an aluminium deck, at the same time as Hayter introduced the aluminium decked Hayterette referred to earlier.

SABO with drive wheels

SABO's first power driven motor mower had a rather unique drive system patented by Sanner in the early 1960s, whereby the operator activated a lever and two belt-driven spiked wheels dropped down engaging with the lawn and immediately propelled the mower forwards.

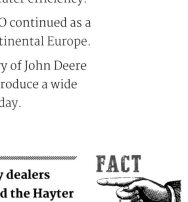

Further developments in 1965 led to SABO manufacturing the world's first fully hydraulic mower for professional use. For the operator, this meant a significant reduction in workload, less wear and tear, more ease of maintenance, and thus far greater efficiency.

Sabo Ad

During the 1970s and 1980s SABO continued as a major force in Germany and continental Europe.

In 1991 SABO became a subsidiary of John Deere and, as with Hayter, SABO still produce a wide range of quality lawn mowers today.

In 1998, garden machinery dealers throughout the entire UK voted the Hayter Harrier 48 the product of the decade.

FACT

A RIDE ON REVOLUTION

It's long been the aspiration of many gardeners to own enough lawn to justify a ride on mower. Being seated whilst attacking the grass, whizzing around the beds with the driving skills required to control a fairground dodgem car, the weekly chore of mowing becomes sheer enjoyment.

Commercial mowers were sat upon from their invention in 1895, and later in the 1920s the trailing seat arrangement was popular; however, it was not until 1953 that E.F. Ranger (Ferring) Ltd of Littlehampton, Sussex, patented the first domestic ride on lawn mower, the 'Ranger Easimow'.

Fitted with a 48cc 4-stroke engine a review in *The Garden* magazine praised the new mower....

Can be driven in reverse.

Epicyclic reduction gear and clutch.

Micro-adjustment for height of cut.

Cutting cylinder removable for grinding in a few minutes.

Lever control for lowering cutter box.

RETAIL PRICE **£68** plus £13 purchase tax

Waterproof sparking plug cover and radio suppressor.

Saddle adjustable vertically and horizontally.

Easimow

Cutting cylinder easily removable in a few minutes.

Fully floating cutter box.

Micro-adjustment of cutter

Quick lift arrangement for cutter box.

Starting lever in carrying position.

Kick starter dogs and drive pulley to cutting cylinder.

Epicyclic reduction gear and clutch.

Twin 7" diameter rollers.

EASIMOW

TRADE MARK

PATENT APPLIED FOR

for the FIRST TIME makes mowing a pleasure

'Revolutions in design are always interesting but when a revolution is staged which is both outstanding in design and completely practical it becomes important. I think both the design and the perfect performance of the 'Easimow' Motor Mower will place it very high in the world of really practical and economic gardening equipment.'

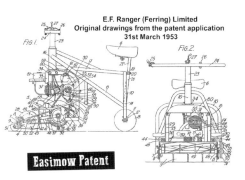

E.F. Ranger (Ferring) Limited
Original drawings from the patent application
31st March 1953

Easimow Patent

Ranger Easimow – Diana Dors

The Easimow also proved to be a hit with blonde bombshell, film actress and singer Diana Dors.

In the late 1950s British Anzani took over production of the Easimow, the company originated in Paris in 1907 established in London 1912. Initially producing engines for the fast growing aviation business, they later made engines for Fraser Nash, AC & Morgan cars. Further diversification then followed with motorcycles,

British Anzani Lawnrider

Anzani Astra Utility

outboard motors, motor cycles and light cars.

There are few surviving Anzani cars, pictured is their Astra Utility, thought to be the only known survivor.

BRITISH ANZANI

LAWNRIDER

A REVOLUTIONARY LAWN-CARE SYSTEM

In the 1960s the Easimow was replaced by the 'Lawnrider' with a sleek typical sixties look, a swan neck alloy frame supporting the comfortable bicycle style saddle. The Lawnrider was available in 18in and 24in cutting widths with a 150cc 4-stroke engine, production continued until the late 1960s.

Advertisements of the time stated that the Lawnrider had enough power to take an 18 stone (114kg) operator uphill too, with the advantage of rolling the lawn at the same time.

LAWNRIDER IS QUICKER *You don't waste time at the end of a strip manoeuvring into position for the return journey. You simply turn through 180 and double back—without stopping, without getting off. The smaller the lawn, the bigger your saving in time and temper.*

LAWNRIDER IS MORE THOROUGH *See how easy it is to mow borders. No more stumbling along with one foot in the flower bed!*

LAWNRIDER CUTS MORE EVENLY *The fully-floating cutter-box maintains a constant cutting height. With a pedestrian mower, varying pressure on the guide handles inevitably causes uneven cutting.*

LAWNRIDER DOES AWAY WITH THE GARDEN ROLLER *Full-weight of machine plus full weight of driver —directly applied.*

DRIVER'S-EYE-VIEW OF SIT DOWN ROLLER-MOWING.

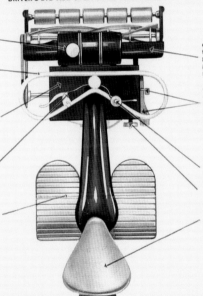

PETROL TANK: Large capacity petrol tank is easy to fill.

HANDLEBARS: Easy steering takes you through end-of-row U-turns, without stopping. You can twist and turn between shrubs, dodge round trees, cut right up along the edge of a border.

ENGINE: A robust 4-stroke engine that's powerful enough to drive **LAWNRIDER** plus an eighteen stone operator—uphill too!

SAFETY CLUTCH: An additional safety measure when starting.

FOOT-REST PLATFORM: Underneath are the rollers. **LAWNRIDER** puts the whole of the driver's weight to work on getting a lawn ironed out.

CUTTER-BOX: The unique fully floating cutter-box can't damage your lawn because it follows lawn-contours independently—won't dig in on bumpy patches.

CUTTER-BOX PEDALS: You lift the cutter-box by stepping on the big pedal, drop it again with the small one. This system lets you take the cutter-box off the ground when crossing paths or riding over ground already covered to empty the grassbox.

THROTTLE: Automatic centrifugal clutch makes starting and stopping easy.

RECOIL STARTER HANDLE

SADDLE: Fully sprung, motor-cycle type. Adjustable for height.

DIMENSIONS

	Without Grass Box		With Grass Box	
	18"	24"	18"	24"
Height	36"	36"	36"	36"
Length	45"	45"	60"	60"
Width	24"	30"	24"	30"

SPECIFICATION

ENGINE 18" model—3 h.p. 24" model—3½ h.p. Powerful four-cycle motor fitted with: Air vane governor, Recoil starter, Large capacity tank, High efficiency silencer, Wet sump lubrication. **TRANSMISSION** Four shoe automatic centrifugal clutch, Epicyclic gear reduction unit, Ferodo lined band clutch, Renolds heavy duty chain, Twin 'V' belt drive for silent and long life. **CUTTER UNIT** Finest quality (Tyzack) Sheffield steel single gate spiral blades. Welded construction for extra strength and rigidity. Special backed-off grinding for efficient cutting and long life. **ROLLER** Three independent die cast rollers ensure high manoeuvrability. **SADDLE** Fully sprung, motor-cycle type, adjustable saddle. **FRAME** One piece cast aluminium with large footrest platform. **CUTTER BOX** Fully floating, will follow lawn contours independently of body of machine hence cannot dig in on bumpy patches. **CUTTER BOX CONTROLS**

Cutter box can be lifted by pressing big pedal. This lifts the cutter box away from the ground so that you can travel without cutting, over paths etc. By depressing the small pedal the cutter box drops down ready for operation. **FOOTREST PLATFORM** Platform conceals the rollers. The **'LAWN-RIDER'** puts the whole of the driver's weight to work in getting the lawn ironed out. **COWLING** Inside the cowling is a robust 4-stroke engine powerful enough to drive the **'LAWNRIDER'** plus an 18 stone operator. **GRASS BOX** Well designed large capacity grass box is easily removable for emptying. **THROTTLE** Single lever control will start and stop machine. Hand clutch is not necessary to use, except as safety measure when starting. **DRIVE WHEEL** Heavy duty wheel 5" wide, bonded with serrated rubber, ensures maximum traction without marking. **WEIGHT** 18" model—190 lbs. 24" model—200 lbs.

A MILLION PANTHERS

The Qualcast Panther was one of the most successful hand mowers ever made. It remained in production from 1932 until the late 1950s with over one million sold; the Panther name was still displayed on later models into the 1970s.

Qualcast Panther - Couple 1936

Panther with family

Qualcast were very clever with marketing, claiming in its trade promotion material that the Panther would 'give the public a guaranteed ball bearing mower at a price lower than many sidewheel machines'. Sidewheel machines were generally less expensive than roller hand mowers.

In a 1959 leaflet they claimed *'Qualcast' and only 'Qualcast' have over 7,000,000 satisfied users.* They had manufactured a range of mowers from the 1920s including a very popular side wheel model 'E' together with motor mowers.

The Panther, probably due to its low price and reliability, appealed to the working class and from early photographs it can be seen that for many families the mower was their pride and joy.

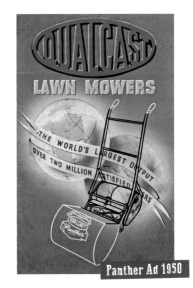

Panther Ad 1950

The 'PANTHER' ROLLER LAWN MOWER

CUTS EDGES AND BORDERS

RETAIL PRICE
£8·6·0

COMPLETE WITH GRASS BOX
All Tax Paid

LARGEST LAWN MOWER MAKERS
QUALCAST
IN THE WORLD

12-inch cut

Panther Ad 1953

1960 TO DATE

PUNCHES, PONIES & COLTS

The Suffolk Punch is a gentle giant and very hard working; these majestic horses helped shape our rural landscape and were vital to Britain's social history. As the UK industrialised and its population grew, the Suffolk Punch pulled the plough, cut the corn and carried the wheat to the mill to feed the towns.

It was no surprise that the Suffolk Iron Foundry of Stowmarket named their first motor mower the 'Suffolk Punch' when it was introduced in 1954. The foundry established in 1920 initially made castings for agricultural and electrical equipment and by 1925 were producing line markers, tennis posts and push lawn mowers.

When the Suffolk Punch was introduced in 1954 it represented more of an evolution than a revolution in mower design. Many of the design features were adopted by other manufacturers in later years and most domestic cylinder mowers still have a similar layout and appearance.

The major components were made from pressed steel and light weight alloys when most of the machines from rival manufacturers were still being built to pre-World War II designs with materials such as cast iron and heavy gauge sheet steel. The Punch was also one of the first popular mowers to be fitted with a compact but

1959 **MARK III**

SUFFOLK PONY

Presents Power Mowing within the reach of all, with many exclusive features of the more expensive models.

12" CUT 27 GNS.
Including Grass Box
TAX PAID

★ Fully self-propelled – Powered by the new Suffolk 80 c.c. two-stroke Engine.
★ Dual Drive – Enabling power to be instantly switched to cutting cylinder only when desired.
★ Automatic Recoil Starter.

The 1959 Mark III Suffolk Pony engine is completely air-cooled – easy and cheap to maintain. Transmission is totally enclosed. The Pony will cut a whole Tennis Court and surrounds in less than half an hour at normal walking speed. Average fuel consumption only one pint of mixture for one and a half hours.

powerful 4-stroke petrol engine. This was manufactured by the company to its own design, unusual in an era when most mowers had either Villiers or JAP engines.

With the Suffolk Punch available in 14in and 17in cutting widths, after three years of production a need was identified for a cheaper, smaller mower and the Suffolk Pony was added to the range.

The Pony with its smaller 12in cutting width had a 2-stroke engine and was being sold in 1959 for 27 guineas compared with the Punch at 37.5 guineas.

For those born after say 1960 we should perhaps explain the 'Guinea'...

The Guinea was originally produced in 1663 and was the largest denomination in British currency until it was replaced by the Pound in 1816. The name 'Guinea' apparently refers to the fact that the gold used for minting coins in the 17th century was supplied by the Africa Company who operated along the Guinea Coast in West Africa.

Since British currency was decimalised on 15th February 1971, the Guinea has no longer been used as legal tender. The term is still used in certain circles such as horse racing to describe values equivalent to one pound and one shilling, or £1.05 in modern currency.

The Pony was only short lived, being replaced by the Suffolk Colt in 1960, retaining the 12in cutting width, but now fitted with the same 4-stroke engine as the Punch.

Did you Know ?

The Colt became one of the most popular small mowers of the 1960s and into the 1970s being light, easily operated and very competitively priced, many are still in regular use today.

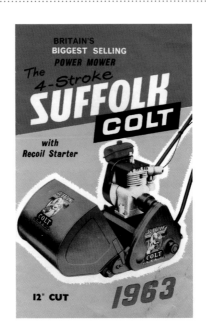

BRITAIN'S
BIGGEST SELLING
POWER MOWER

The
4-Stroke

SUFFOLK
COLT

with
Recoil Starter

12" CUT

1963

IT FLOATS ON AIR!

Christopher Cockerell (1910-1999), an English engineer, invented the hovercraft in the mid 1950s. The prototype SR-N1 made its first trip across the English Channel on 25 July 1959.

Hovercraft SR-N1 1959

The hovercraft was a revolution in sea travel. In the 1960s the fleet of craft could be seen plying the English Channel and The Solent between Portsmouth and Ryde and Southampton and Cowes.

However, the passenger hovercraft were hit by the rise in fuel prices in the 1970s and the Hovertravel service between Southsea and Ryde is now the only passenger hovercraft service left in Britain.

In 1963, Karl Dahlman from Sweden had adapted Cockerell's principle to work on a lawn mower after initially experimenting with an engine on a dustbin lid; a British patent was granted that year.

Flymo 1960s Logo

In 1965, the Flymo Company was founded and the first hover lawn mower was produced at their factory at Newton Aycliffe, County Durham, England.

Which? magazine was quick to test this revolutionary mower and in August 1965 declared it Best Buy Power Rotary

FACT

The hovercraft began its life in 1955 when inventor Sir Christopher Cockerell tested out his idea for a floating/flying craft by putting a cat food tin inside a coffee tin. He then blew a jet of air through the gap between the two tins to create a cushion of air and the hovercraft was born.

Mower at £35.14s despite close competition from the Hayter Hayterette.

Whilst all Flymos were initially made with petrol engines, in 1969 the company became part of the Electrolux group and the world's first electric hover mower was produced, note the early models were blue.

The mower was not initially a great seller perhaps due to its plastic construction which, although it appeared flimsy, the material was the same as that used to make police riot shields.

Toro Flymo Poster 1965

THE MOWER THAT FLOATS ON AIR

FLYMO
BY
TORO

Introduced in 1965

To increase business the company utilised door to door salesmen with the intention of demonstrating the mower and its ease of use to the lady of the house!

At the same time a survey of several thousand housewives was completed including asking 'what colour they would prefer the mower to be?', the unanimous answer was 'ORANGE' and that's what it's been from 1977 to this day.

Dinner's at eight, and you're cutting it fine. Everything's just right. An hour to go, the table set, the wine chilled – and you've made sure you're looking good.

The finishing touch? A perfect lawn to look out over as the sun goes down. At the simple flick of a switch your Flymo electric airborne rotary mower will float over your grass on a cushion of air, banks and slopes, over level paving stones, under bushes, right up to the patio edge – in no time at all, without a trace of effort.

So little effort, in fact, that when eight arrives you and your lawn are both looking good – as cool and fresh as the spring evening.

Flymo Ad 1975

Bond Bug

Space Hopper

General Lee

But why orange you may ask? Well those of a certain age will remember that it was the 'in' colour for interior design during the early 1970s with cushions, curtains, carpets all in orange hues. Others will remember the Bond Bug, Space Hopper and General Lee (Dukes of Hazard 1969 Dodge Charger)!

Flymo Orange

Shortly after the change in colour, Flymo patented the world's first grass collecting hover mower in 1978. This was a major achievement as it had to blow down to create the hover, at the same time sucking the grass cutting upwards.

Flymo adverting in 1980s stated
'It's a lot less bovver with a hover'

Qualcast countered when marketing their best-selling Concorde electric mower by saying 'It's a lot less bovver than a hover' (they sold over 6 million).

FACT

MOWERS, FASHION AND FLAIRS!

Fashion has often been used in the marketing of lawn mowers, as early as the 1870s Follows & Bate of Manchester illustrated their croquet mower being operated by a young lady in very fine clothes, this being continued by Thomas Green & Son of Leeds in the early 1900s.

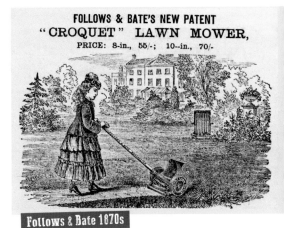

FOLLOWS & BATE'S NEW PATENT "CROQUET" LAWN MOWER,
PRICE: 8-in., 55/-; 10-in., 70/-

Follows & Bate 1870s

GREEN'S MOWERS & ROLLERS
WORLD RENOWNED

By Appointment to H.M. the King.

EVERYONE who takes a pride and interest in the possession of a beautiful velvety lawn should become the owner of a Green's Lawn Mower. This scientifically perfect machine cuts cleanly and closely, is remarkably light and easy to use, quickly adjusted and always in smooth running order, while the name "Green's" is sufficient guarantee of the high quality and exceptional finish.

BRITISH MADE

Interesting List No. 9 Free on Request.

Thomas Green & Son 1900s

The 1950's American mower manufacturer Fairbanks-Morse linked up with Coles of California famous at that time for supplying and designing clothes for Hollywood and swimwear for Christian Dior; in a 1956 mower advertisement there was no sign of grass but two very fashionable women!

In the 1960s Hayter were keen to depict women in smart, relaxed clothing operating their mowers with their 1966 catalogue depicting more female than male operators of their various machines.

Hayter 1960s

Atco 1920s

Atco 1950s

Atco in 1922 had depicted a lady dressed in a rather smart woollen suit handling their 22in motor mower which weighed in at 308 pounds (140 kilos), and by the 1970s they were certainly keeping up with the fashion world with dungarees, flairs and tight jumpers as seen in the accompanying images.

Atco 1970s

RACING MOWERS

Lawn mower racing was started back in 1973 by an Irishman called Jim Gavin, who, with a group of friends had gone down to The Cricketers Arms in Wisborough Green, West Sussex, for a few pints one lunchtime to discuss his latest motorsport idea.

Jim was heavily involved in rallying and like all motorsport at this time, sponsorship was becoming more common. Jim didn't really like this and wanted to create a form of motorsport that didn't involve lots of money and was readily accessible to everyone. As the pints flowed they looked out across the village green and there was the groundsman mowing the cricket pitch. It was then they realised that everyone had a lawn mower in their garden shed so they said 'let's race them', and they did! They announced there would be a race in Murphy's field and on the day about eighty mowers turned up!

The main objectives were and still are, no sponsorship, no commercialism, no cash prizes and no modifying of engines. The idea being, it would keep costs down and resulted in lawn mower racing being described by *Motor Sport News* as 'the cheapest

form of motorsport in the UK'. The BLMRA still sticks to its origins as a non-profit making organisation; any profits are given to charities or good causes.

Lawn Mower Racing takes place all over the country from Wales to Norfolk and Yorkshire to Sussex, appearing at Country Shows, Fayres and Steam Rallies.

Racing generally starts in May through to October, incorporating The British Championship. We also have The World Championships,

The British Grand Prix, The Endurance Championship and the most famous of all, The 12-Hour Endurance Race.

Over the years lawn mower racing has attracted motor racing legends and celebrities. Sir Stirling Moss has won both the British Grand Prix and the annual 12-Hour Race. Derek Bell, five times Le Mans winner and twice World Sports Car Champion, has won the 12-Hour twice and one of those was with Stirling. The actor Oliver Reed, who lived in Sussex, regularly entered a team. The BLMRA is mentioned in the *Guinness Book of Records* with the fastest mower over a set distance and the longest distance travelled in twelve hours. Other famous names who have been seen in the paddock are Murray Walker, Alan deCadenet, John Barnard (Ferrari F1 designer), Phil Tuffnell, Jason Gillespie, Chris Evans, Guy Martin and Karl Harris (British Super Bike riders).

RADIO CONTROL & ROBOTS

With the growth of technology in the mid-20th century inventors and electrical engineers were keen to look at ways of automating all types of machines and appliances. Jim Walker from Portland, Oregon, USA, is credited with inventing the first radio controlled lawn mower in 1948. He adapted a petrol engine side-wheel mower with clutches operated by solenoids controlling the power to each driving wheel.

Jim Walker

The mower would run straight ahead, but either wheel could be declutched by radio to steer the robot or turn it around. The engine and cutting cylinder would run at a constant speed and a triggering bar at the front shorts out the spark plug if the mower bumps into an obstruction.

In the UK at the 1959 Chelsea flower show, H.C. Webb & Co., an established mower manufacturer, launched their radio controlled mower. The mower was displayed as part of 'The Times Garden of Tomorrow', for *Dr Who* fans the mower does perhaps resemble K9!

Webb Radio

Webb Radio

The cutting width was 14in and powered by a $1/3^{rd}$ HP 24-volt motor running at 1,450rpm. The operating range was stated as being up to one mile; bearing in mind remote video had not yet been invented, I am sure it could only be used successfully within eyesight.

The first mass production robot mower was introduced in 1995 by Husqvarna, a Swedish company, powered by solar panels. Later developed to operate from an on battery making it possible to mow the lawn after dark. Robot mowers are usually kept within the lawn area by perimeter wires either laid on the surface or buried beneath the ground. With the advancement of technology they can now check the weather, height of grass and return to their docking point to recharge the battery when required.

Did you know

With mobile technology you can now relax on the beach in Spain whilst controlling the mowing of your lawn in Surrey using a mobile app.

Today's top of the range Husqvarna robot mower, with a retail price of £3,800, has many features including built-in sensors using ultrasonic technology to help the mower detect objects by lowering the speed to avoid hard collisions. That should keep the local cats happy! In addition, it can cope with forty-five degree slopes, onboard GPS to create a map of the garden plus full surveillance from smart phone, tablet or laptop. An alarm system is incorporated enabling the mower to be tracked in the event of theft, or perhaps should it wander off to pastures new.

Husqvarna Robot Mower

Robot mowers have now been with us for over twenty years but I feel they have a long way to go before extinguishing the traditional mower and the satisfaction it brings.

THE MOW-CYCLE

Whilst we would perhaps consider the hybrid a new idea, Lloyd Lawrence & Co., USA, featured an image of a hybrid mower/bicycle in their 1894 brochure. This was to promote their new sidewheel mower 'The Chicago' which featured a cylinder that could be removed without taking the frame apart.

Lloyd Lawrence & Co., London, was formed in 1877 and imported various hardware including mowers from the USA. The company is still trading today under the name of Lloyds & Co. of Letchworth.

Chicago Lloyd's Bike

LLOYD'S CHICAGO
AMERICAN LAWN MOWER
☆ ☆ ☆ ☆
SOLE CONSIGNEES,
LLOYD, LAWRENCE & Cº
29, Worship Street, LONDON, E.C.

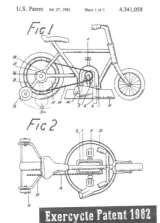

U.S. Patent Jul. 27, 1982   Sheet 1 of 3   4,341,058

Fig 1

Fig 2

Exercycle Patent 1982

In 1982 a patent was granted in the USA for 'Exercycle Mower Apparatus' which was basically a bicycle with additional stabiliser wheels, fitted with a rotary mower driven by the pedal power. The patent emphasised the green credentials of the machine, no pollution from fossil fuels, less noise pollution and the benefit of free exercise for the operator.

Sunkyong Cycle Mower

A production mow-cycle was made in 1985, the Sunkyong Bicycle Mower, produced by a mail-order company known for its esoteric and extravagant gadgets. A full two-page review was made in an American cyclist magazine in 1986 including a full specification.

Price was $399.50, weight 63lbs, also available in Dichondra Green and by coincidence distributed by a company based in Chicago!

My Mower Cycle

TURF ART AND THE RHS

The lawn has always been a particularly English concept with its neatness and definitive parallel stripes giving an air of status, whilst perhaps in the average household it has developed into more of a utility area for games, dogs and barbeques; many large houses, stately homes and parks retain large expanses of lawn.

Garden designers and turf managers are always looking to add something new and exciting to their projects, in 2013 the then turf manager at the Royal Horticultural Society's much loved and extensive

garden at Wisley, Surrey, was experimenting with 'turf art', the vast areas of lawn being a blank canvas. To carry out this task a mower with a narrow cutting width is required, with no new mowers being suitable The Budding Foundation Museum of Gardening was consulted.

During the late 19th and early 20th centuries most mower manufacturers supplied small mowers with cutting widths of 6in which were used for finishing tight areas where the horse-pulled mower could not reach.

A selection of mowers from the museum suitably sharpened and prepared are now on permanent loan to the RHS and the turf art has been developed further in recent years by James

Bourne (Turf Manager) and his team. The mowers range from a Ransome's 6in Automaton Minor from around 1905 to a JP Mini Mower from the 1960s.

In 2017 our mowers were used by the Wisley team to create designs on the lawns of The Royal Hospital, Chelsea, in readiness for the Chelsea flower show.

Other parks and gardens are looking to add interest in a similar way after observing the artwork at Wisley and we are busy preparing additional mowers to help them achieve some eye-catching designs.

MY SEARCH FOR THE FIRST 'MAN THAT WENT TO MOW'

MY SEARCH FOR THE FIRST 'MAN THAT WENT TO MOW'

When my initial interest in lawn mowers as a collector was aroused, I delved around on the web and found a few promising websites, sent a couple of emails, but was met with a wall of silence from most. Was I trying to enter a mysterious cloak-and-dagger world? A welcoming response was received from The Old Lawn Mower Club; I had now found 500 like-minded people worldwide who shared my new interest.

OLMC Members

The club was formed in 1990 when a small group of enthusiasts first got to know each other including the founding member, Keith Wootton, who says: *At the time we thought there might be a 'few people' who were interested in lawn mowers and their history. But after a short while it was clear that there were many more people out there and the club soon became established.*

Grassbox magazine

I signed up in 2010, quickly adopting the appealing dress code of a well-oiled boiler suit together with hat of choice. Eight years on I am actively involved as a committee member, in contact with collectors worldwide and regularly writing in the club's quarterly 'Grassbox' magazine.

Whilst most members are content collecting, restoring and showing their mowers, I wanted to know more about this unique piece of machinery that had such an impact on social history and indeed who was behind it?

A considerable amount of research was completed with the generous help of the Museum of English Rural Life in Reading, Berkshire; their archives are extensive with extremely helpful staff who have readily assisted me on my visits to the reading room and promptly responded to my many emails seeking information.

My research has uncovered an unacknowledged genius of an inventor in Edwin Beard Budding; this book will tell more of his story, his inventions and how his lawn mower has progressed over the years.

To personally engage with Edwin Budding I decided to visit the Stroud Valley, Gloucestershire, where he spent his life and also look to see if I could find any remaining family, as I have had an interest in genealogy since the early 1970s when I researched my own family history.

Mr and Mrs Pedersen 1898

Prior to the first visit I had obtained a map of St Michael's Churchyard in Dursley where EBB was buried, so on arriving I had an idea where to search. Whilst the map indicated his grave was around eight plots in from the boundary wall, I could find no sign. I engaged the help of a man walking his dog through the churchyard and whilst he could not help, he did show me the grave of Mikael Pedersen (1855- 1829) the Danish inventor of the Pedersen Bicycle. Though never hugely popular, they enjoy a devoted following and are still produced today. Their unusual frame design was marketed as cantilever, and features a distinctive hammock-style saddle. Variations include lightweight racing, tandem, and folding designs. Other Pedersen innovations include two- and three-speed internally geared rear hubs.

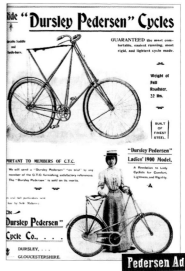

Pedersen Ad

Unfortunately Pedersen lacked business acumen and was both profligate and prone to being cheated. He left Dursley unannounced when in his sixties, leaving his family behind.

Pedersen was spotted by a friend selling matches in London, and the friend arranged to pay his way back to Denmark in 1920. He died in 1929, poor and virtually unknown, and was buried in an unmarked grave in Bispebjerg, a suburb of Copenhagen.

In 1995 a collection was started by enthusiasts for the Pedersen bicycle to raise funds in order to bring Mikael Pedersen's remains back to Dursley and re-bury them there. This was achieved in 1995, and the service was attended by over 300 people including the Bishop of Gloucester, representatives from the Danish Embassy and Pedersen's grandchildren.

Continuing my search I wondered if the churchyard was a resting place for inventors, if so where was Budding? If Pedersen had all this attention for a marginal modification to the bicycle no doubt I should be looking for a sepulchre or mausoleum containing Budding's remains?

Another glance at the map of the churchyard and pacing out the measurements it became clear that the perimeter boundary had changed and part of the burial ground had sprouted a terrace of modern houses! At last I had found him, a large flat slab, hard up against the boundary fence covered with brambles and moss.

How lucky the developers had not taken an extra metre, we may have lost him!

Removing the brambles, and gently easing away some moss the inscription was revealed as follows....

TO THE MEMORY OF EDWIN BEARD BUDDING
OF THIS TOWN. MACHINIST WHO DEPARTED THIS LIFE
THE 25TH DAY OF SEPTEMBER 1846 IN THE 50TH
YEAR OF HIS LIFE
IN AFFECTIONATE REMEMBRANCE OF
ELIZABETH BUDDING OF THIS TOWN WHO DIED
24TH OF JUNE 1874. AGED 80 YEARS.
CAROLINE BUDDING, SECOND DAUGHTER OF THE ABOVE
WHO DIED FEBRUARY 21ST 1898. AGED 72 YEARS.
ALSO OF FRANCES ANNE. DAUGHTER OF THE ABOVE
AND WIFE OF JAMES NEWTH, CABINET MAKER OF THIS TOWN
WHO DIED 11TH NOVEMBER 1898. AGED 76 YEARS.

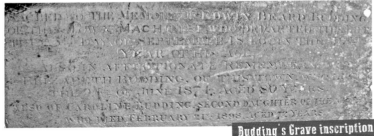

Budding's Grave inscription

Unlike Pedersen there was no reference to the great inventor's revolutionary invention, just a simple slab marking his resting place. I have seen videos of Elvis Presley fans paying tribute to their hero when visiting his tombstone, strumming guitars and humming 'Love me tender', but how could I honour my hero? Well, being a lawn mower historian, I always have an old mower in the back of my truck, just for the odd emergency if I come across some grass that needs trimming, what could be better I paid tribute to Edwin Beard Budding by cutting the grass around his grave with a 100 year old Thomas Green & Son 8in 'Silens Messor' lawn mower.

I was thinking whilst mowing, what other mowers had passed within six feet of their creator since he was laid to rest in 1846, hopefully an original Budding, perhaps later a Ferrabee, Ransome's, and no doubt the popular Silens Messor which was manufactured from 1859 to the 1930s.

Job done, I headed off to find the site of the former Phoenix Iron Works, in Thrupp, just outside Stroud, this being the place Budding & Ferrabee produced the first ever lawn mower. On arrival at a rather quiet small industrial estate with a modern type two-storey office block occupying the site of the former Iron Works. Not particularly exciting, but within

Stroud Brewery
presents

BUDDING

CAMRA'S CHAMPION BEER
OF GLOUCESTERSHIRE
2006 & 2008

CHAMPION BEER
of Gloucestershire
2006 & 2008

STROUD
Brewery

Available on
draught and
in bottles

Award-winning
beers brewed with
Cotswold-grown barley

Ask for our beers at
your local, or buy direct
www.stroudbrewery.co.uk
07891 995878

BUDDING
Stroud Brewery
Pale Ale
4.5%

a few metres on the opposite side was Stroud Brewery, established here in 2006. Just two months after opening, it launched its first brew, Budding Pale Ale (4.5% ABV), this was awarded Champion Beer of Gloucestershire at the Cotswold Beer Festival and again in 2008. The business has grown steadily and now employs over twenty people.

I returned to the brewery in April 2015, to attend a small informal event where a blue plaque to commemorate Edwin Budding was unveiled by David Withers, President of Ransomes Jacobsen. The plaque having been provided by Chris Biddle on behalf of the British garden machinery industry. Chris has produced magazines for the turf and mower industry for over twenty-eight years; perhaps owing his livelihood to Budding.

Stroud's Museum in the Park, was next port of call. Set in the beautiful grounds of Stratford Park in a Grade II listed

Budding Plaque unveiling - David Withers

17th Century wool merchant's house; their collection tells the fascinating story of Stroud District's rich and diverse history.

With over 4,000 objects on display, including dinosaur bones, historical paintings and, I am pleased to say, the story of Edwin Budding, including two early lawn mowers manufactured by Ferrabee in the 1850s after Budding's death. My old friend Mikael Pedersen also gets a mention with an example of his bicycle on display.

Having now trodden in Budding's steps and indeed over him, I decided to focus on his family to see if any living descendants could be located; with the aid of the internet you can quite quickly establish a family tree. It soon became clear that Edwin Budding's direct line would be small as he only had three children, with only two grandchildren from his son Brice Henry Beard Budding.

I did find some interesting information on Brice, he was an assistant engineer 2nd class in the Royal Navy having served on HMS *Fury* and HMS *Pearl*, at the age of twenty-eight in 1858 he was 5ft 11½in with light brown hair and blue eyes. This detail of information is not usually found, but it appears Brice jumped ship in Calcutta, on 26th October 1858 and the Inspector General of Police in Sydney posted a notice in March 1859 offering a £3 reward for a 'Seaman Deserter' from HMS *Pearl*.

Livingstone's 'Ma Robert'

Of perhaps greater interest is that when HMS *Pearl* left Liverpool for Calcutta on 10 March 1858 with Brice Budding on board, a passenger was

Dr David Livingstone together with Livingstone's paddle boat *Ma Robert* which was strapped to the deck for Livingstone's second Zambesi expedition; they arrived at the Zambesi delta on 14 May 1858. I can only presume that Brice would have met Dr David Livingstone during the two-month journey on board, I wonder if he mentioned in conversation that his late father had invented the lawn mower?

In those times it was quite common for sailors to jump ship and join the better paid Merchant Navy; we can only assume that Brice was successful in getting a ship back home to England as he appears in later census records.

Further research on the family has revealed that only two direct lines of descendants remain, only one of which retains the 'Budding' name. I have regular contact with both lines and had the pleasure of meeting John Budding in 2014 who is a direct great, great grandson of Edwin Beard Budding. When we met I arranged for John and his wife Joan to stand with a replica of Budding's original mower on loan from a close friend and fellow collector Colin Stone.

John and Joan Budding

Subsequently I have now met with Alex Budding, John's grandson born in 1992, the great, great, great, great grandson of Edwin Budding. Alex has a keen interest in the family and can be seen here also with a replica of his ancestor's lawn mower from 1830.

Alex Budding

Fake images of Edwin Beard Budding

The direct descending families have no known memorabilia relating to their now famous ancestor, I was therefore puzzled as to how the internet had various images purporting to be portraits or photographs of Edwin Budding. At the time of his death in 1846 he was a little known engineer living in a small village, the camera was in its infancy and it would have been most unlikely that he would have had a portrait or photograph taken. Thorough research has proved that the internet images are to use President Trump's favourite words, 'FAKES'.

Charles James Cockerill

The most often used image is in fact taken from a portrait of Charles James Cockerill (1787–1837) a British entrepreneur, the image was placed on a French website blog in 2009 with 'Portrait de Edwin Beard Budding' beneath it, the result being it was then adopted by many people and organisations who were writing about the lawn mower's invention. It may be that Cockerill and Budding were mentioned on the same area of the web due to their connection with the woollen industry, with search engine results being misinterpreted.

The second most popular fake is Sir Edwin Arnold (1832-1904) an English poet and journalist, the only connection with Edwin Budding is the shared Christian name!

Finally we have Rudolph Chambers Lehmann (1856-1929) an English writer and Liberal Party politician, the only

Sir Edwin Arnold

evidence here appears to be reference information on the web mentioning that the Lehmann brothers were budding cricketers!

Rudolph C. Lehmann

I have had some success in persuading website owners and bloggers to remove the images and help avoid distorting history for future generations, but we are unfortunately in a world where many people believe what they see on the world wide web without question.

Perhaps in another one hundred years Pedersen will be credited with inventing the lawn mower and Budding the bicycle, after all they do have close links, including presently residing only a few metres from each other!

Did you know?

Just two months after opening, Stroud Brewery launched its first brew, Budding Pale Ale (4.5% ABV), this was awarded Champion Beer of Gloucestershire at the Cotswold Beer Festival and again in 2008.

BIRTH OF THE BUDDING FOUNDATION

During my childhood my parents created an environment of helping others, although poor in income and assets, they always had time to help neighbours, friends and relatives, sometimes complete strangers. I like to think I have inherited these traits and have adopted them when I could throughout my life.

On reaching fifty years of age and being totally disillusioned with the world of banking, its policy of profits before people, bullying of staff and customers, I decided to look for a change and perhaps something more rewarding for me, others and the environment.

With a keen interest in horticulture, conservation, together with good DIY skills, I initially embarked on improving my CV with some related qualifications as my 3 CSEs and one GCSE from school plus thirty-five years in banking would perhaps not help with my new direction. Initially volunteering with the South Downs Rangers on conservation projects I met interesting, real people some of whom had perhaps been neglected and trodden on by society. I also embarked on a Pond Warden's course where a chance meeting would lead to spending the next eight years regularly volunteering at local schools.

The meeting came about at Sussex Wildlife's Woods Mill nature reserve in Henfield, West Sussex. Whilst taking a lunch break from the pond studies a woman came in to our room with a plastic box seeking advice; she had a small naked baby pigeon in the box which had fallen from its nest. In the absence of any reserve staff I did offer suitable advice. In conversation I established I was speaking to Pippa Bird, headmistress of Manor Hall Primary School in Southwick, who, on hearing I was on a pond-related course, immediately requested advice on their school pond.

This led to a most rewarding relationship with the school, initially creating a large new pond and conservation area, to include reed beds and a dipping platform. In the following years an environmental project building a bottle greenhouse, together with a regular gardening club. A fantastic achievement came in 2010, when

Bottle Greenhouse

Budding the Inventor to inspire and help Budding young people

The Budding Foundation was registered with the Charity Commission in January 2014, many fund raising events followed – quizzes, sponsored walks, attendance at fetes, steam rallies, talks and the opening of a museum in March 2016 which has become its focal point. Now over £20,000 per annum is being raised.

The beneficiaries are many and varied, young people aged five to twenty-one who have lacked support at some stage in their lives and need a helping hand with funds to improve their future.

School Club

the children won two awards in a national sunflower growing competition, competing against 800 other schools they grew the tallest plant and largest flower.

Working with the children was inspiring often seeing them develop in areas that the restricted school curriculum would not have achieved, and with friends and family working with deprived children, both in social services and local children's homes, I decided I wanted to focus my efforts to help support young people.

To celebrate my 60th birthday in 2013 we organised a fund raising event Clivefest , this being a 12 hour beer and music festival, attended by many friends and relatives. Musicians from far and wide gave up their time to help raise funds, including Asbo Derek, The Bandana Collective, Pog, Atilla the Stockbroker and Tracey Curtis who made a 500 mile round trip from Wales for a special performance.

Raising money for good causes is not easy, if you are giving up a great deal of time it needs to be enjoyable. Many people link fund raising to a personal interest, sporty types run marathons, accomplished cooks make cakes, companies and businesses link fund raising to their particular industry. I was now involved in horticulture and becoming a lawn mower historian, what better than dedicating a charity to Edwin Beard Budding.

THE MUSEUM EMERGES

THE MUSEUM EMERGES

It's the dream of many collectors to perhaps create a museum dedicated to their particular interest but is usually cost prohibitive as substantial funding would be required, with many existing museums failing due to insufficient resources to continue. I was aware that Tates Garden Centre in Hassocks, West Sussex, housed the South Downs Heritage Centre within their complex and were planning a major redevelopment.

I feel it appropriate at this juncture to give a brief history of the Tate family as follows...

Records can trace the Tate family living and working on the land in the Sussex area as early as 1601. By 1782, William Tate was known in the Findon area primarily as a timber merchant, with Thomas and James Tate working as blacksmiths and Sarah Tate as a dressmaker. By the mid-1800s, the main trade was as wheelwrights and master carpenters. George, and then his sons James and George Junior, would work with oak in the summer, and ash, beech and elm in the winter to produce cart and coach wheels for the local area.

By the 1870s, the Tate family had extended into Portslade where Thomas Tate was landlord of the Clarendon Arms pub and James Tate worked as both a butcher and a shopkeeper. During the late 1800s Alfred Tate started a laundry on the corner of Foredown Drive and Old Shoreham Road in Portslade. Most laundries were transporting their goods by horse and cart at the time, but after a car accident in 1896 Alfred converted a Daimler to what is regarded to be the first commercial vehicle

in Sussex, using the cart top of a horse drawn vehicle to achieve this conversion. The business was ultimately taken over by his sons Albert and Fred, who started to operate a film transportation business. To keep their film distribution vehicles in shape, they built a small corrugated shed alongside the Southern Cross laundry to carry out simple repairs on their vans. They were soon helping to fix others vehicles. Word grew that this was the place you could take your vehicle to be repaired, adding another arm to their business. The Tate brothers expanded their workshop as more and more customers came to see the Tate brothers for parts and repairs. In 1929 the Brothers purchased a corner site at Southern Cross in Portslade which now operates as head office for the family's motor business.

The outbreak of World War II led to the car business taking a back seat while the family concentrated on engineering activities that would help the war effort. Being close *to the coast and a port town, the Tate Brothers became actively involved in ship salvage. After the war demand for cars surged, and suddenly Albert found both the engineering and the car firm very busy, so he approached his son John to see if he would run the car business. A young, new member of the family coming in to an old established business brought with it pioneering changes.*

In 1982, horticulturist Jonathan Tate joined the family business and started about establishing a plant nursery in mid-Sussex propagating unusual plants for sale to members of the public. Inspired by the new concept of the garden centre emerging in the UK, where people could visit and purchase everything they needed for their garden from one place, he set about developing what is now Tates of Sussex Garden Centres.

I made my initial approach to Jonathan Tate in late 2013 to see if there was any possibility of involvement with the new Heritage Centre, resulting in meetings with Jonathan and his daughter Sarah who was joining the family business to project manage the new development.

With the old heritage centre demolished and exhibits safely stored, a new 1,500 square metre vernacular style Sussex barn was erected, constructed from restoration-grade oak fashioned by hand in a traditional workshop by a team of master carpenters. Whilst the museum would cover all aspects of gardening, being dedicated to Edwin Budding it would bring his story to life and the onward development of the lawn mower. The design and layout of the museum was given very careful consideration; experience of similar collections was that the exhibits are often low down hidden behind barriers with little information being available, our brief was to get them 'in your face' so to speak enabling the visitor to engage more with the exhibits.

Another priority was to ensure the displays gave as much information as possible: patents, posters, old advertisements etc. to inform and educate the audience.

There was much experimentation with display plinths and units, all constructed by the talented in-house carpentry team headed up by Wayne and Terry, who ensured our ideas were turned into reality.

The large barn also accommodates the Sussex Food Hall, with an extensive stock of local produce, and of course plenty of Budding Pale Ale. Original and contemporary art is expertly displayed by local artists Martin & Di Nee in the Cube Gallery. Sarah Brangwyn 'Made & Making' provides a selection of workshops and classes, from sewing and knitting to floristry and paper craft.

In March 2016, the Heritage Centre and Museum opened, the official opening ceremony followed in July with the ribbon being cut by Jim Buttress. Jim was awarded the Royal Horticultural Society's Victoria Medal of Honour (VMH) in 2006 the highest accolade possible in the gardening world for his unrivalled contribution to horticulture and has become a valued supporter of the charity and museum.

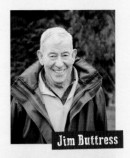

Jim Buttress

Official opening ceremony

The museum now entering its third year, is continually changing with increasing visitor numbers and perhaps sets the scene of how small museums need to link in with larger complexes to survive in this financially challenging world.

INDEX